Innovation and Society

**Smart Innovation Set**

coordinated by
Dimitri Uzunidis

Volume 15

# Innovation and Society

Françoise Cros

iSTE

WILEY

First published 2018 in Great Britain and the United States by ISTE Ltd and John Wiley & Sons, Inc.



ISTE Ltd
27-37 St George's Road
London SW19 4EU
UK

www.iste.co.uk

John Wiley & Sons, Inc.
111 River Street
Hoboken, NJ 07030
USA

www.wiley.com



Library of Congress Control Number: 2017962219

British Library Cataloguing-in-Publication Data
A CIP record for this book is available from the British Library
ISBN 978-1-78630-203-8

# Contents

# Introduction

Education and training are implicated in all society's important questions, to the extent that, in France, school has become a focus for social tensions and difficulties such as drugs, authority, respect for others, violence, harassment, religious faith, sustainable development, social and ethnic inequalities.

Is it not said that every society gets the school that it deserves?

In fact, education is deeply entwined with our uncertainty faced with a future whose shape we can only see poorly. And yet, we never cease to want the best for our children, and so we work towards this goal by providing everything that we consider good for them, with a deep desire to intervene in the present so as to accomplish this worthy wish. The present is interpreted as offering possibilities for the future, and school is thus the melting pot for interventions among the youth, still malleable and subject to the beliefs and intentions of adults[1]. Adults believe that they can work with these possibilities and inscribe their dreams on them: it is from this that pedagogical innovations emerge, as containers without specific content[2].

---

1 Did not Durkheim say that education is "the action exerted by adult generations on those who are not yet ripe for social life? It has for its object to arouse and develop in the child a certain number of physical and mental states which political society in its entirety, and the social milieu to which he or she is particularly destined, demand from him or her" (*Education and Sociology*, PUF, 3rd edition, p. 51, 1992).

2 By this I understand that the momentum towards a better future represented by innovation often relegates what provoked it to the background.

In a way, the world of schooling is imbued with a passion: that of wanting to change the present, bend it towards a very different society to come, always considered the bearer of a brighter future than the present, that the adults will open before the first steps of children.

This is the issue with schooling; we cannot observe it without tainting our observation with the urge to reform it. However, this observation has changed over the years, and it is clear that more than a century ago, school was a luxury for many and represented the hope of social and cultural upward mobility. The corollary of this was the development of an elite, or rather a caste, which was very much linked to the social level of the family. Now that schooling is compulsory, it has become more banal, the source of growing criticism, and considered a consumer good. It is hard to see schools once more being closed to the majority for the benefit of a minority (although…) and we will see how this is reflected in the facts, notably through the prism of innovation.

And yet, make no mistake: when we survey the landscape of schooling, many great changes have taken place. This does not, however, mean the disappearance of social division, which indeed seems more insidious in its process, while being more visible and felt less clearly. It is often described as counter-intuitive, an involuntary outcome, the result of the implementation of certain innovations which have in practice had a preferential impact on already-favored student populations. However, as some would say: school changes! Again, we must examine how this happens, the place and the role of the "factory" of innovation in transformation[3].

For more than half a century, in France, after having invaded the territory of the economy and of technology, innovation has conquered education, including schools. The exponential use of the term to designate different activities has often confused matters. We must at least recognize that using the term allows the user to find themselves in a position to create something new, to reply to a need emanating from multiple sources, a need which is felt and must be filled. Innovation is on the whole considered positive – with magical consequences.

3 On this subject, see the works of Antoine Prost, notably *Du changement de l'école : les réformes de l'éducation de 1976 à nos jours,* Le Seuil, Paris, 2013.

According to intellectuals, our current society appears confronted with a multiplicity of movements and changes, faced with the acceleration of time (have you noticed how our young people speak more quickly than their parents and how the immediacy of Twitter entertains their thought?).

If innovation at school is invoked more and more as a way of escaping the difficulties linked to the complex objectives and purposes of school, it nonetheless remains the case that this innovation has never really benefited from serious work on its meaning, its sense and its impact. Is this not primarily due to the new identity that contemporary schools have taken on? How would we define an ideal school so as to achieve a consensus?

And yet, this is not for lack of focusing on the problem and suggesting all sorts of remedies, or rather informed advice![4]

Innovation is most often invoked in business in conjunction with a creative spirit, which is able to develop new products for the presumed greater benefit of the company to which it belongs. It is thus understood as the original appearance of a tangible object, often perfected by technology. To understand this form of innovation, therefore, we only need to follow the trajectory of the emergence of the prototype up to its testing and its generalization, a trajectory which is certainly complex, but which is difficult to follow in the case of social innovation, the category to which innovation in schools belongs.

Over the last 10 years, innovation has conquered the economic domain, faced with a world of investors who seek consistent financial profit from new production. The permeability of innovation with regard to the ecosystem has rendered it frugal – that is, less costly in energy, requiring the marketing of products which are less energy-hungry, while still improving their performance. The often-criticized consumer society produces innovations which are more ecologically friendly, while still being more useful (for example, electric cars). Is innovation, then, no more than an adaptive response to social transformation?

4 Two bodies were commissioned to make proposals on rescuing schooling from its ghetto and opening it towards a learning society: the Taddei committee, which concentrates on the place of school in a learning society, and the National Council of Innovation for educational success. The two reports were published simultaneously in March 2017.

Innovation has also been – somewhat less valued in its infancy than technical innovation – introduced to the associative social network in the form of social innovation. This takes the form of a perceived response to unmet social needs such as health, the aging of the population or food supplies. It seems that innovation at school is closer to the nature of social innovation than technical innovation. We can then ask why, among the many social innovations studied through reports and publications, innovation at school is so seldom mentioned. What might be so specific and particular about innovation in education that it fails to excite debate?

What justifies such caution? Given that words become worn out more quickly than their content, this is doubtless due to its longevity compared to more recent words such as educational transition or pedagogical experimentation or alternative schooling, in various registers. In the end, it is enough to give an unchanging social activity a new name for it to become something different (at least, for those with little memory). There is thus some aspect of educational transition which proves its value in the very goal of innovation: to create a more just world through schooling. Utopia[5]?

What differentiates Freinet's pedagogy from a contemporary innovative pedagogy based on written correspondence organized between students at two distant schools? Is it the fact of using tablets or computers? Sending mail to correspondents on the other side of the world via the Internet? What constitutes the quiddity of innovation in education, especially in pedagogy, where after centuries the fundamental question remains: how does a small human being learn, and how do they develop socially and intellectually? Does the explanation lie in the unceasingly changing context, and – if so – do the multiple questions at the heart of education remain the same in the face of sometimes contrasting ideologies, convinced of their righteousness? However, revisiting the pedagogies of certain precursors, considered by many to be obsolete and outdated, may be a salutary reminder for certain pedagogies which claim to be innovative. Did Durkheim not write:

5 *Les Cahiers Pédagogiques* are not wrong; the link between society and school is complex. Is that organization's motto not: "Changing school to change society; changing society to change school"?

> "I believe that it is only in studying the past that we become able to anticipate the future and to understand the present, and that, consequently, a history of teaching is the best of pedagogical studies"[6].

This book attempts to shed light on this question by laying out the origins, directions and issues of innovation when occurring in schools, including in the pedagogical relationship.

To do this, we will discuss the socio-historical context of innovation while explaining it in the context of school. We will then survey the different levels and fields of innovation (from the organizational to the personalized pedagogical relationship), to lead to an analysis of innovation occurring in public or private schools, with its political and ideological consequences.

The conclusion of this work will take the form of a question: in the end, what makes us want to innovate, without going backwards, amid the strong politicization of new activities, given the current profusion of alternative schools with different pedagogies, which seem at first sight to possess the same conceptual ingredients and rhetorical register, and which, fundamentally, in their actual organization, obey different social ideologies? What is the meaning of the growing role of foundations, of alternative schools and of sponsorship, faced with schools which declare themselves to be innovative? Is this a sign of a neoliberalism which dares not speak its name in schools? The enthusiasm with which politicians take schooling and proposals for innovation hostage shows the extent to which no-one has any interest in clarifying this notion, and what it hides in terms of implications and explanations. Is innovation therefore a gimmick for amusing the spectators, or a necessity for the real evolution of schools?

6 DURKHEIM E., *L'évolution pédagogique en France*, PUF, Paris, p. 16, 1938.

> "I believe that it is only in studying the past that we become able to anticipate the future and to understand the present, and [illegible]

# 1

# Innovation in Socio-Historical Context

## 1.1. Meaning and significance of the word "innovation"

### 1.1.1. *The origins of innovation*

Innovation is part of today's habitual way of speaking which consists of projection into the future, that is, transformation of the present time to invite individuals to become "project animals": teaching people to do projects is our contemporary redemption. Thus, "innovative" projects abound, furnished with the best assets in their supposedly beneficial aspirations. Their establishment and inevitable decay, up to disappearance, resulting from their confrontation with complex social reality, are filled with contradictions.

However, innovation has not always been synonymous with development regarded as positive, linear and rational. From the Middle Ages up until the 16th Century, innovation was seen as disruptive, transgressive, disturbing, even Satanic. It was urgent to maintain institutions in their current state, and to struggle against attacks linked to any desire for transformation. Respect for institutions such as the Church was fundamental. Whoever wanted to introduce something new was condemned as a heretic. We must understand that there were two kinds of innovation: those carried out for the material benefit of people, regarding which religions had contrasting positions, even internally; and those which innovated in religious practices themselves, for example, saying the Catholic mass in the vernacular language instead of Latin. It is thus impossible to state that religions have always been firmly opposed to innovation; they have often subtly played with it in order to reinforce their identity and their contemporaneity.

Michel de Montaigne hated innovation: "Nothing oppresses a state," he wrote, "except innovation; change only brings about injustice and tyranny." In his *Essays*, innovation is synonymous with "novelty", a word which the author used in a pejorative sense.

*The Quarrel of the Ancients and the Moderns* came from the need for solid and stable models, as society was perceived as intrinsically fragile and the fear of returning to primeval chaos was strong. Diderot himself wrote: "all innovation is to be feared in government." The meaning of the term innovation was, in that era, apocalyptic, in strong contrast with the modern flavor of the term. This *Quarrel* illustrates the tension between these two extremes.

The rise in power of science and technology has led us to a positive connotation of the term, linked to examples of inventions as brilliant as they are useful. "From the start of the 19th Century, innovation became the god that we still worship today."[1]

As the taste for innovation became stronger, the definition of the word became radicalized to the point of excluding tradition and imitation. Innovation thus took over in all domains, including that of painting, where modern art appeared, or in literature, surrealism. A kind of "terrorism" of innovation seems to have gradually appeared, emerging from a blank slate. Girard would go as far as to speak of "self-deception". He raised the question of the very important role of imitation, which seemed relegated to the background or even to disappear, in the process of learning[2]. He added:

> "When the humility inherent to the master-disciple relationship is seen as humiliation, the transmission of the past becomes difficult, or even impossible."

1 GIRARD R., *La voix méconnue du réel. Une théorie des mythes archaïques et modernes*, Grasset, Paris, 2002.

2 WINIKAMEN F., *Apprendre en imitant ?*, PUF, Paris, 2015 (republication).

In other words: innovation, imitation and tradition are components of the same process of social evolution.

### 1.1.2. *Innovation as the inverse of tradition*

There is no innovation without tradition, that is to say, without the roots which fertilize innovation, invite it to build on the old to create the new, to give it meaning. Was it not Paul Ricoeur who said:

> "Tradition in the singular means transmission of things said, of beliefs professed, of norms assumed. It is thus first of all to cross a temporal or cultural distance. Now, this crossing is only possible if tradition remains half of the couple which it forms with innovation. In my previous work, I referred several times to this tradition-innovation couple, tradition representing the side of the debt owed to the past and the reminder that nothing comes from nothing."[3]

We must add here that tradition is endlessly revisited by those who consider it, including for pedagogical innovation, through pedagogues of the past such as Pestalozzi, Montessori, Steiner or Freinet. It is doubtless through contemporary reading, often unconsciously, of the great pedagogues of the past that innovation, the other side of the coin, appears. Moreover, as Ricoeur specifies in the same work, this is not so much in what might have happened than in the unfulfilled promises of the past. In addition, this past is part of a future for those who lived in it, in what Koselleck[4] calls the "past future".

In fact, innovation frequently emerges from dissatisfaction felt at the present, thus becoming a possible response to overcome this dissatisfaction. Now, the latter does not arise from nothing, but is part of the momentum of a reality which must be surpassed. This momentum is the product of a knowledge of the past and its possible futures which were not able to come to pass. In this sense, pedagogical innovations are strictly linked with the past experiences, sometimes in an unconscious manner, of their initiators, if

---

3 RICŒUR P., "Identité narrative et communauté historique", *Cahier de Politique Autrement*, October 1994.

4 KOSSELECK R., *Le futur passé. Contribution à la sémantique des temps historiques*, EHESS, Paris, 1990.

we consider that a society's relationship to its past experiences and to its hopes conditions its possibilities of action. Koselleck adds:

> "Throughout modern times, the difference between experience and expectation never ceases to grow or, more exactly, modern times only experience themselves as new times from the moment when expectations, in their impatience, become more and more remote from all previous experience." (p. 324)

All this is linked to the fact that innovation cannot dispense with time, particularly with what Bachelard[5] describes as psychic time: "duration is only a number whose unit is the moment." It is "a dust of moments, more, a group of points which are grouped together more or less closely by a phenomenon of perspective". It is doubtless because time is seen as the present moment that the past is mishandled, when it is consubstantial with creation and innovation!

### 1.1.3. *Innovation versus invention and creation*

Nietzsche wrote that "only that which has no history can be defined". Now, although innovation is indeed a concept which, since the 13th Century, has been used extensively in many societies, it is our current era that uses and abuses it to designate exponentially a collection of activities considered to be economically as well as socially beneficial.

First of all, three words are often confused with innovation and have a common point of involving the new: invention, creation and innovation.

Invention has marked our history: throughout time, human beings have sought to increase their power to act on their environment and thus to acquire ever more powerful intellectual and material means[6]. It consists of an act often seen as the fruit of a single person to whom we give the name of "inventor", even if, when we look from further away, this invention is closely linked to its context. Another word is very close to it: discovery and, for example, discussions around the discovery of the printing press by Gutenberg bear witness to this. In fact, although the inventor and the

---

5 BACHELARD G., *L'Intuition de l'instant*, Stock, Paris, 1932.
6 LEROI-GOURHAN A., *L'homme et la machine. Evolution et techniques*, Albin Michel, Paris, 1943.

discoverer are clearly designated and often glorified, they did not spring out of thin air, but in a social context which allowed their emergence. Simply put, until recently, context has not been taken into account in the forms of thinking about society.

We find this in historical perspectives, which were once stories about great men and the great events which marked them out, while now history is centrally focused on the daily lives of people[7]. Now, concerning Gutenberg, it is said that other individuals of the same era were on the cusp of inventing the printing press! Galileo in 1623 evokes the inventor of the telescope, making him out to be an opportunist in a situation where its components were at hand and a possible new usage was conceived:

> "We are certain that the inventor of the telescope was a simple optician who, manipulating by chance various shapes of glass, also happened by chance to look through two of them, one convex, the other concave, held at different distances from his eyes. He realized with astonishment the effect they produced, and from there, discovered the instrument."

The inventor is often convinced that his idea, formulated as a dream or a utopia, will be called upon to become a founding practice.

Schumpeter himself distinguished invention *(art of insight)* and innovation *(the carrying out of combinations).* The first is exogenous with respect to the field, whereas the second, innovation, is an endogenous factor, a process that arises from the field and only has meaning in its socialization: "Invention is an exogenous factor from the endogenous character of innovation."

Innovation is thus an invention which has succeeded, in the sense that it becomes integrated into people's social practice towards a goal which subsequently becomes institutionalized, that is, becomes part of the habits of people and is thus considered more as innovation. In innovation (which some call *rupture* to signify that it renews certain ideas), *it is the modes and habits of consumption and of life which change.* For example, the electric toothbrush was invented around the 1940s, but it was only 20 years

7 On this theme, see the works of Michelle Perrot on women and those of Arlette Farge on archives.

afterwards that it became an innovation, through the process of progressive trivialization of its use.

The path would be that shown in Figure 1.1.

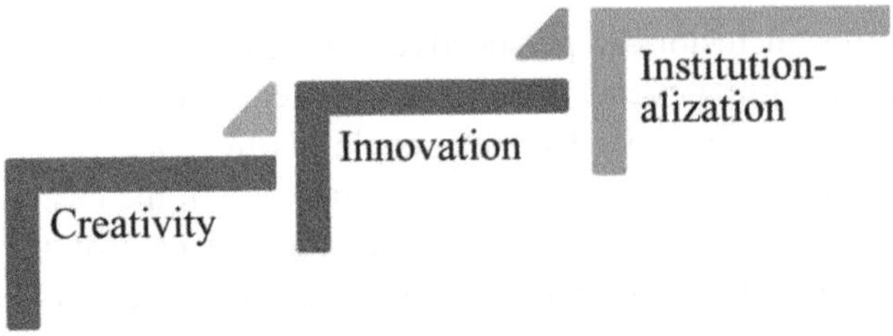


**Figure 1.1.** *Process of development of innovation*

As for creation, it arises from a very strong personal component, motivated very little by the production of a technological object aimed at improving practical life. D. Anzieu[8] distinguishes two components of creation: one linked to the psychic work of creation, the relationship of the author to the work, which he calls poietic: the other, named aesthetic, which concerns the fate of the work, the relationship between the work and the public, including its social recognition.

The poietic is envisaged as a creative process obeying general laws which the author attempts to release through experiences and works already completed, and he marks out "five stages of the path of creation" (these stages are not linear; they can be zigzag, spiral or irregular):

– inspiration consisting of an interior apprehension, in the same way that we may be seized by cold;

– return to consciousness in a sort of regression;

– the central core around which the work will materialize and be constructed;

– the accomplishment in detail of this invention, that is, composition in its proper sense;

– the creator makes the difficult, arbitrary decision that their work is finished and ready for public view.

8 ANZIEU D., *Le corps de l'œuvre*, NRF, Paris, 1981.

Freud[9] discusses the multiplicity of parameters (sexual, social, familial, psychic, etc.), which are involved in all creative processes. The creator is seized, it is not him or her who seizes, by a profound force that mocks reason and consciousness. Let us not forget that the Greek word for creation is *demiurgia!* Freud adds, concerning Leonardo da Vinci:

> "[there is] a special disposition – though about its determinants (which are probably organic) scarcely anything is yet known. [...] We consider it probable that an instinct like this of excessive strength was already active in the subject's earliest childhood. Amongst other things it involves sublimation, that is, the state of changing its immediate goal for others, not sexual, possibly placed higher on the scale of values."

However, Anzieu explains that the process of creation can also be collective. But, from his point of view, this collective creation is rather made up of individual creations (and he gives the example of the surrealist painters whose mutual influences are undeniable, while each is still credited with their own creation). Consequently, innovation is not a creation.

### 1.1.4. *Innovation and creativity: a brief incursion into school*

The creative process is rare and not to be confused with creativity, with which all individuals are blessed, if not everyone expresses it. The difference between them comes from the "launch", this interior apprehension.

Moreover, Anzieu certainly distinguishes creativity and creation when he writes:

> "So-called minor arts such as decoration, gastronomy, weaving, pottery, etc., demand creativity[…] this is an area in which there is not considered to be creation. So, we speak of innovation instead of creation."

Does using the word innovation instead of creation (especially in the arts) indicate enlarging the field and the number of individuals who may be designated creators? Does the word innovation therefore become a higher

---

9 FREUD S., *Leonardo da Vinci and A Memory of His Childhood*, Gallimard, Paris, 1987.

category, encompassing all the words whose meaning is related to novelty? This leads Anzieu to say that, currently, nothing becomes worn-out more quickly than the new!

Creativity is thus a capacity that each of us possesses; it consists of the personal dispositions and opportunities which induce creative behaviors. There exist many training tests and exercises which allow us to develop this capacity. These range from brainstorming to schemes for developing ideas, along the lines of: "I'm going to a desert island and I can only bring five things: what, and why?" Similarly, Bono's six-hat exercise which teaches us to sequence thought in six distinct domains, each represented by a different-colored hat, is another example (white for neutrality in the quest for facts, red for the emotional part, green for creativity, black for pessimism, yellow for optimism and blue for the organization of ideas).

Creativity is equally the result of three indispensable criteria: expertise (you must be familiar with and possess knowledge in the domain where this creativity will develop); being motivated to change things, and have divergent thinking which often creates relationships between thoughts or things which are usually perceived separately, with no relationship (what might be called the competences or capacities of reorganization), as shown in Figure 1.2.

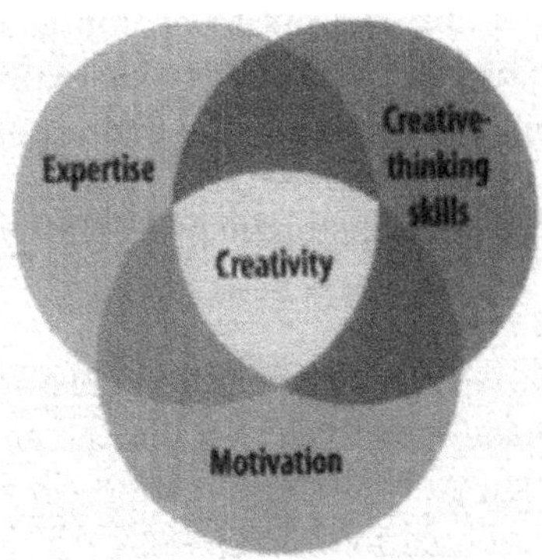


**Figure 1.2.** *Determinants of creativity (source: Innovation, Education and Training in the EU Member States, 2009*[10]*)*

10 Report edited by FERRARI A., CACHIA R., PUNIE Y. and published by the European Union.

This capacity, to produce content which is both new and adapted to the constraints of the situation or the task to be performed, is enacted in three phases:

– the emergence of an idea (of the order of insight);

– the development of this idea, since we all have original ideas, but they remain unused or censored;

– the deployment of the idea, which may furthermore be taken up and deepened by what we may call a change agent who is none other than the person who had the idea.

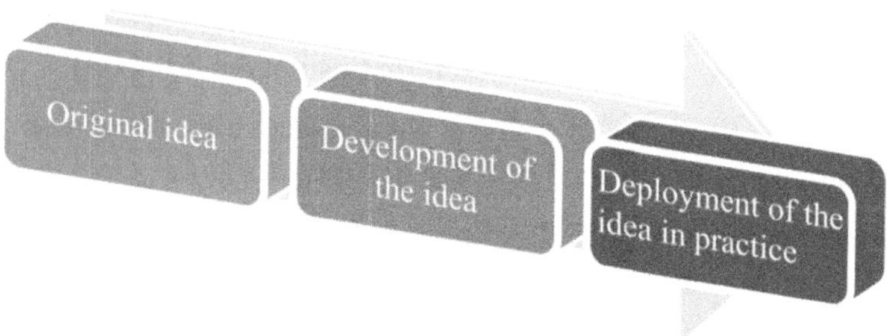


**Figure 1.3.** *The three phases of implementation of creativity*

Creativity becomes the starting point of innovation, revealing itself to others who may adopt it or draw inspiration from it. In firms, there are people known as "change agents", on the lookout for these original ideas for which they have the foresight to see the possibilities. An innovator launching an innovation all on their own is rare at present, even in schools where work is increasingly carried out in teams. However, the role of change agent does not officially exist in middle schools or high schools. This is a person who will spontaneously support an idea as it is put into action (which may be done by the head of a school or by a leader of the teaching staff). According to Frascati's manual, "innovation is an idea transformed into something that can be sold"; doubtless, this is the stage at which a change agent may intervene.

In the field of schooling, creativity thus becomes a competence which the students must also acquire. Many new reforms of the educational system insist on the acquisition of this competence as a fundamental step. To integrate with contemporary society, the teacher is thus required to provide activities for students to develop their own creativity. They are led to develop what Puozzo calls "a pedagogy of creativity"[11].

Researchers have studied the ways in which schools might encourage the development of the competence of creativity among students, in particular in music courses.

Thus, Giglio[12] identified five dimensions, which indicate ways in which teachers may take charge of developing this competence among students and, on that basis, create possible modules of professional training. These five dimensions are:

– the explicit formalization of a requirement to develop creativity among students in programs for educational reform (institutional support);

– the ability of teachers to envision creative activities, or representations of this creativity by teachers (knowledge of expectations);

– creative activity itself in its role as a source of creativity, supported at the same time by consciousness of creating, in a practice of creativity linked with research and development;

– necessary social interaction as a facilitator of development for this competence (creativity is based on collaborative and emulative activity at school);

– forms of contribution in setting up creative activities in the content of teaching and the use of new tools, such as digital devices. Training teachers to attain the goal of developing the competence of creativity among students cannot be done without considering these five dimensions.

As said previously, every individual possesses the potential for creativity. They can seek it out and develop it, or leave it lying fallow. The reasons for such an attitude are multiple; however, current society increasingly encourages every individual to be creative, if only to cope with rapid social and technical change. Each person is encouraged to be creative in their working lives.

---

11 PUOZZO I. "Pédagogie de la créativité: de l'émotion à l'apprentissage", *Éducation et socialisation*, 33 | 2013, published online 1 September 2013, accessed 1 June 2017. See http://edso.revues.org/174 (DOI: 10.4000/edso.174).

12 GIGLIO M., "5 dimensions pour étudier le changement de l'éducation du maître pour améliorer l'apprentissage créatif", *Journal for Educators, Teachers and Trainers*, vol. 5, no. 1, pp. 80–89, 2013.

It is also in this sense that creativity becomes a priority in the development of the competences of students in education, to face the uncertainties of the 21st Century society[13]. There are tests which provide us with indications of a person's potential to develop their creativity and to innovate. The two main examples are those of Torrance[14] and those of Guilford[15], based on convergent thought and divergent thought, the latter being conducive to creative activity.

There are, however, ways to increase this capacity for creativity, which is based on dispositions such as openness to new experiences, cognitive and affective variables such as perseverance, self-esteem or a tendency to active participation in social networks. It is clear that if we want to develop this competence in schools, the role of teachers will have to change to allow for free expression by students, while at the same time showing them paths to perfection.

Certain pedagogic practices are based on inviting artists into classrooms, who can encourage more open and original behaviors without substituting themselves for the teacher.

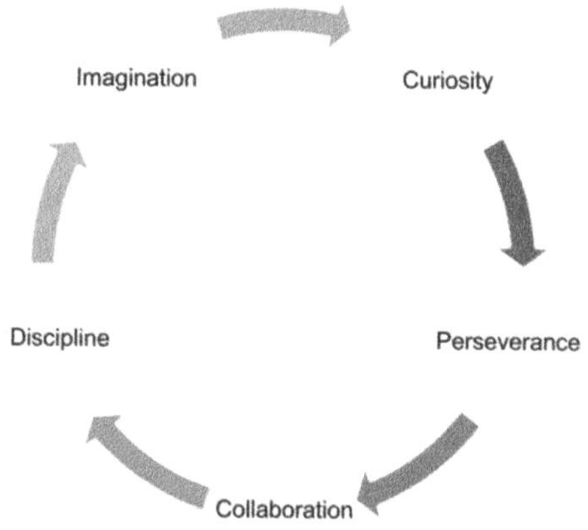


**Figure 1.4.** *The five components of creativity*

13 Miller R., Looney J., Siemens G., *Assessment competency: knowing what you know and learning analytics; it's time for a breakthrough, Promethean Thinking Deep*, Research Paper no. 3, Promethean Education Strategy Group, Blackburn, United Kingdom, 2011.
14 Torrance E.P., *Torrance Tests of Creative Thinking: technical-norms manual*, Research Edition, Princeton Press, Lexington, Massachusetts, 1966.
15 Guilford J.P., *Characteristics of Creativity, State Office of the Superintendent of Public Instruction*, Gifted Children Section, Springfleld, Illinois, 1973.

Figure 1.4 sums up the five possible components of creativity.

Creativity has been widely studied in the United States, and two authors have carried out research in this area: Robert J. Stenberg and Tod Lubart. They define creativity as follows: "the capacity to produce work which is at once new (that is, original and unexpected) and appropriate (that is, useful in relation to the constraints of real work)"[16].

Creativity is doubtless the central quality of all innovation[17]; without it, it is impossible to escape from habitual patterns of thinking, which Watzlawick describes as type 2[18] change (*thinking outside the box*). However, every person can be creative without necessarily being innovative, meaning that even though innovation needs creativity, the latter will not necessarily become innovation. The latter demands a more complex process of social change, composed of negotiation and creation of networks – that is, of strong social capacities.

## 1.2. Innovation in the world of beliefs and values

If there is a whole hidden area at present on the subject of innovation, it is that of religion. This becomes an important question in the current context of the exponential development of currents based on old or new religions. The etymology of religion, based on the meaning of "relinking" people to each other in a "communion" of shared "beliefs", of faith, leads to a link with innovation as the aspiration for a better future world (a sort of Parousia).

The parallel between innovation and religion, in a kind of abandonment of physical time for eschatological time, seems hidden among current positivist reflections on innovation. A hot topic if there ever was one!

Yet, the famous study performed by Max Weber on the sociology of religions marked the beginning of reflection on this link between religion and innovation. This study tried to show a possible link between religious doctrines and the innovation-producing structure of the Western economy. Weber thus designed a plan of comparative study of religions and economies

16 STERNBERG R., LUBART T, *Handbook of Creativity*, Cambridge University, 1999.

17 On this topic: "Vers une éducation plus innovante et créative", *Dossier d'actualité, veille et analyse*, no. 70, Institut français de l'éducation, Lyon, 2012.

18 WTAZLAWICK P., WEAKLAND J., FISH R., *Changements: paradoxes et psychothérapie*, Le Seuil, Paris, 1975.

and their mutual influence. The constitution of an "ideal type" allowed him to focus on the multiple and contradictory components of liberal capitalism as it existed in that era – that is, an economic system in search of profit, of ever-greater profitability, enslaved to the order of the market, based on a rationalization of speculation which summarized for him "the spirit of capitalism". American civilization was his reference point, notably the writings of Benjamin Franklin which reflected this particular ethic. In other words, Weber offered a comparison between Protestantism and Catholicism, stressing that the first, by the very fact of its principles, allows innovation, in contrast to the second.

In fact, Protestantism is based on the redemption of sins (indulgences). Thus, each person may take charge of their own life's project, leading towards an individualism which is little seen in Catholicism, where the community takes precedence. Moreover, Protestantism invites each of the faithful to read the Bible for themselves, to make their own interpretations of it and to discuss them with others (which implies everyone knowing how to read). On the other hand, the Catholic Church is based on the interpretive filter of priests who tell the faithful what to think, in a sort of collective submission.

Furthermore, Protestantism does not have the same relationship to money, which can be used in market transactions, while the Roman church imposes a vow of poverty and of everything pertaining to finances, which seems to link it to other religions such as Islam. The condemnation by the church of usury is based on biblical texts such as the famous Deuteronomy 23:20-1, where usury is declared lawful for the enemies of the faith. Of all merchants, the usurer is the most accursed, because he sells something given by God and thus sells something which does not belong to him. Time is given to man, and he may not construct it and claim to dominate it[19].

This somewhat brief comparison allowed Weber to conclude that the troubling proximity of the outbreak of innovations in Germany, for example, was linked to the relationship with the world dictated by the dominant religion, while in France, the Church hindered the development of innovation and commercial exchange. However, at the same time, we might observe commercial and innovative development during the Renaissance in

19 "You shall not lend at interest to your brother, whether it be a loan of money or of food, or of anything for which interest is demanded. To the foreigner may you lend at interest.", *Deuteronomy, École biblique de Jérusalem*, 23, Éditions du Cerf, Paris, pp. 20–21, 1956.

Florentine society. This latter was of course Catholic, which somewhat contradicts Weber's intuitions. However, he can be credited with posing the question of links between innovation and religion.

As a result of his illustrious work, many debates have taken place and still continue. J.M. Berthelot[20] analyzed Weber's explanatory structure to take account of the relationship between Protestantism and capitalism:

> "This relationship contains not two but three levels – that of homologies between Protestant ethics and the spirit of capitalism, that of the transformation of religious asceticism into the Puritan way of life, and finally that of the structural effect of a change in behavior of certain actors within a given tradition. Protestantism is an ethic and the matrix of a new ethos whose secularized reflection we find in the spirit of capitalism."

However, Weber strikes a tone closely linked to the conception of innovation as situated in a liberal capitalist market. He writes[21]:

> "The first innovator encounters mistrust... How can we not see that only his uncommon strength of character enables the new-style entrepreneur to keep his composure? [...] It is only by virtue of ethical qualities which are well-determined and strongly developed that he also finds the ability to inspire his clients and workers with absolute confidence in his innovations." (p. 71)

He reserves a place for Judaism which, due to its history, is based upon insecurity. Due to this, for him, "Judaism stands on the side of venture capitalism and speculation; in a word, its ethos is that of a capitalism of pariahs" (p. 201). We can easily see here the image of the hero of capitalism found in the work of the founder of innovation within capitalism, namely Joseph Schumpeter.

However, Max Weber's thesis was strongly debated at the time and found some opponents. For example, H. Karl Fisher considered Weber's book to be an overdetermined text, showing confusion between the spirit of capitalism and German investment, in the sense of this openness to

20 BERTHELOT J.M., *L'intelligence du social*, PUF, Paris, p. 33, 1990.
21 WEBER M., *The Protestant Ethic and the Spirit of Capitalism*, Gallimard, Paris, 2004.

innovation. In fact, it pertains to the German population's idea of having a profession, one based on the feeling of duty fulfilled in its very exercise. The latter is thus "the chief content of a person's moral action", and money is a means of exchange par excellence. Thus, we better understand that the transfer of feelings operates precisely on the field of value added to money. Issues of religion do not come into it. And this author adds that the Renaissance capitalists, in Genoa, Florence or Venice, demonstrate that, in the medieval era, the capitalist spirit was also able to exist within Catholicism – even if we remember the strong presence of Savonarola and his school. There is thus a parallelism to be drawn between the spirit of capitalism and religion, but not causality; the latter must be found elsewhere.

Such statements, though, only bind their author. For others, they are not solidly founded and only reinforce Weber's thesis on the idea of a particular capitalist ethos. M. Gauchet[22] also expresses reservations as to the unilateral and obvious link between the "spirit of capitalism" and "ascetic Protestantism", and considers that it is "the leading edge and the emergent breakthrough of an immense subterranean movement which, little by little, imposed the active optimization of the terrestrial realm in place of the old restrictive submission to the intangible".

The arrival of Lutheran and Calvinist principles enabled behaviors more consistent with the options of capitalism, in particular allowing financial usury, "making money work", something unthinkable in some religions where the vow of poverty and rejection of financial interests are basic principles. Conversely, the Catholic Church was a brake on overt innovations, eager to exploit financial resources as a motor of innovation.

The resistance of the Catholic Church envisaged by Weber allows him to use this explanation. It is true that the image of the Catholic Church of that era was synonymous with the maintenance of tradition and the structural inability of its ecclesiastical society to renew itself, ignorant of industrial society and its effects, and organized around the mainly rural world, as existed before 1870. Did not Denis Diderot say that "when the priest favors innovation, it is bad; when he opposes it, it is good"?

The papal encyclical *Rerum Novarum*, concerning the condition of the working class, 1871, rejected any liberal doctrine while still being mindful of

---

22 GAUCHET M., *Le désenchantement du monde*, Gallimard, Paris, p. 100, 1965.

the conditions of workers. Funnily enough, this text emphasizes that social equality is not a good thing to the extent that inequality is beneficial to the body politic, allowing it to fulfill the set of functions necessary for civil society. Such an approach runs up against the innovator invoked by Weber whose very basis is equality, to the extent that every single person can innovate in an open market society. For the Catholic Church, everyone has their place in society, even – and maybe especially – when it is unequal, this inequality not being felt as an injustice, but as what must be. This inevitable and necessary inequality is reinforced by the condition of man after the original sin.

This link between innovation and religion is above all seen as something demiurgic, leading us to view innovation at the same time with fascination, attraction, fear and revulsion. Does Greek mythology not speak of Prometheus being punished for having stolen fire from God for men[23]?

In *Prometheus Bound* by Aeschylus, Prometheus responds to the coryphaeus that it is based on fire that men will learn many arts from him. There is in innovation a kind of messianism that sees progress[24] as the completion of creation. These two elements – technical modernity and arrangement into capitalist societies – proceed from the simplifications of Max Weber. Moreover, it is sufficient to note the extent to which the Catholic religion, even if some religious groups still refuse modernity, especially in the conduct of the liturgies, has been affected by it through, for example, the electrification of churches. A religion can remain doctrinally correct while still benefiting from technical advances, that is, from innovations.

However, this idea of a link between innovation and religion is not exhausted, and some of our contemporaries try to find explanations for it not in religion itself, but in the differentiated practices which are manifested within it. Researchers on this question are convinced of a link between religious practices and the economic and technical development of a society. We are no longer truly dealing with religion, but with religiosity.

23 LAGREE M., *La bénédiction de Prométhée: religion et technologie*, Fayard, Paris, 1999.
24 We will deal separately with the question of innovation and progress. See section 1.4.

A preliminary survey was carried out[25] in several countries to compare them to each other in 2000. Religiosity was defined as an individual position with regard to 11 parameters measuring attitudes towards openness or refusal of innovation and five parameters identifying the degree of religiosity (belief in God, importance of religion, importance of God in their life, perceived expectations of the Church). Figure 1.5. plots the macroresults against each other according to country, as a function of two opposed parameters (strong or weak belief in religion, strong or weak attitude towards the development of innovation).

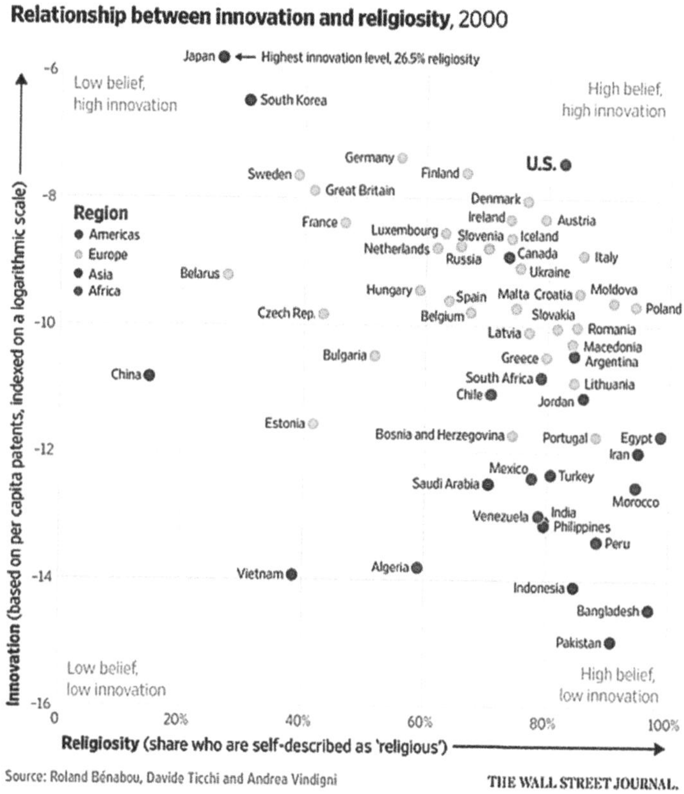


**Figure 1.5.** *Relationship between innovation and religiosity in several countries. For a color version of the figure, see www.iste.co.uk/cros/innovation.zip*

25 This survey was carried out by three American researchers: BÉNABOU R., TICCHI D., VINDIGNI A., "Religion and innovation", NBER, *Working papers*, no. 21052, March 2015a.

Figure 1.5 shows, for example, that the United States shows high religiosity and belief, while being open to innovation. There is therefore a paradox if we cling to the hypothesis that religiosity (rather than belonging to a religion) is an obstacle to the development of innovation. The study carried out by these three American researchers in their own country is based on this aporia[26].

First of all, they divide the set of countries into three types of society: secular societies, including Western Europe; theocracy, defined as a very religious society, refusing knowledge which does not come from the Church; and a society between the two previous types, such as the American system which combines the two with strong inequality between individuals, creating a "right to religion" and requiring rich and poor to coexist faced with discoveries and ideas, in a sort of indifference.

What interested these researchers was to go into more detail about understanding the concrete practices implicating the individual in their religion. However, it was difficult for them to identify the actual practices; they therefore constructed a broad sample of Americans using the country's demographic studies and taking individual statements with respect to 11 measures of (declared) attitudes to what they considered to be openness and five measures of religiosity. From the multitude of data, they distinguished three global attitudes:

– an attitude towards science and technology;

– an attitude towards new ideas, change and risk-taking;

– qualities envisaged for the education of the young in order to encourage (or not) children to develop their imagination, their independence, their determination and their perseverance.

The results of this study showed that the links are complicated and that, for example, in Florida, the mixture of attitudes displayed made an extension of innovation possible; similarly, the social level of individuals is a strong determinant linked to religious practices and thus does not allow us to say if religiosity or social class determines openness to innovation.

26 BÉNABOU R., TICCHI D., VINDIGNI A., "Forbidden fruits: The political economy of science, religion, and growth", NBER, *Working papers*, no. 21105, April 2015b.

Let us include the religion of Islam, whose sometimes strict doctrines recall a saying of the Prophet[27]: "Take care of invented things, because all these things are an innovation, every innovation is a bewilderment and every bewilderment leads to hell." Innovation is contrary to the Sunna, that is the Islamic law of prophetic tradition. Moreover, the word "innovation" does not exist as such in the Arabic language; the word *bid'ah* is often used, which means both a new idea and a heresy.

In other words, if we can say that certain doctrinaire religious sects are an obstacle to any innovation, for the others, the results are uncertain. However, let us not forget that there is no innovation, particularly social and in education, without a background of values and a vision of the future which guides its direction. It is because the innovator has an image of what the world of the future will be after the enactment of the innovation, that they work on it with an enthusiasm which sometimes borders on militancy. Note, however, that education's status as the quintessential place where values or belief in a better or redemptive society are invested often results in a close link between innovation in education and religious belief.

## 1.3. Innovation in schools: what is its definition?

For more than 20 years, we have studied innovation and, more precisely, a type of innovation which is not often celebrated because it is difficult to identify: innovation in education, innovation in schools or pedagogic innovation[28].

Yet, there is no shortage of definitions of innovation: we have collected over 600 of them and all of them are relevant, to the extent that the word is mentioned relating to new activities[29], activities which increasingly define our contemporary societies.

What differs is not so much the definition itself of the word, as the social use made of it and, in this case, the meaning arising from this contextualized

27 See the site www.islammessages.com, accessed May 2017.

28 We will examine the difference between these innovations further on.

29 As an example, we provide here the definition posted on the Pollen site of the Ministry of Agricultural Education: "Innovation in education is a deliberate process of transformation of practice by the introduction of pedagogic and organizational novelty which is the subject of dissemination and which aims at the sustainable improvement of the educational success of pupils or students", March 2017.

usage. For example, to speak of innovation in political discourse signifies that the user wishes to mark their intention to change things in a way that they consider to be positive, or again, to speak of innovation for a prison shows a desire to improve the performance of the institution and to avoid accidents, without necessarily making its inmates the beneficiaries of such a transformation.

All these usages blur the meaning of the word. Moreover, it is interesting to note that, in public interventions, we have given a precise definition of the notion of "innovation at school" and yet, at the end of every presentation, people have come to see us and noted that we had not given a clear definition of school innovation; and this has systematically reoccurred!

What explains such a refusal to listen? Are there words which are undefinable because of their social significance and their implicit issues? Is it not the case that, through a contamination effect, the word innovation, because of its lability and continuous interrogation, suffers from being caged in a definition which is stabilized once and for all?

In any case, this attitude is not based on a lack of definition, but on the deep desire that there should not be one in particular. Why? Two explanations are possible. The first is based on the fact that an innovation is defined by the results of the innovative activity, and not by a semantically stable concept, thus allowing openings in active interpretation. The second is the refusal of a framework: innovation is envisaged as the freedom to do what is intuitively better, refusing *a priori* analysis and explanation. The act is thus considered to be impossible to formulate, only visible and comprehensible in the breadth of professional activities. The innovator is not accountable to some censor or legislator for the nature of this new practice[30]. As Nietzsche said, a definable word is a word without a history, meaning that words belong to the eras and situations in which they are used.

Yet, when we study the plethora of innovations, certain stable and common ingredients emerge, invariant features which, taken together, account for innovation in education. There are five of them, in no particular order, which we will explain and analyze in their relevance.

30 For more on this idea, see the following work: JOAS H., *La créativité de l'agir*, Éditions du Cerf, Paris, 1999.

First of all, innovation, following from its etymology, indicates something new. This latter is introduced into a precise and situated context. However, what qualifies as new? What indicates that something is new? In the case of a technological object, the question is scarcely relevant, as it is enough to look at it to understand that previously there was nothing like it, and therefore it is new. Yet, we must not forget that every innovation, even of a product, is only innovation when it is the result of social appropriation, that is, of new social behaviors being adopted.

However, in the context of education, what can we say is new? Is it setting up multi-age classrooms? Was that not already done in the 19th Century in the "mutual instruction" system? Peer learning has been used for a long time. Getting an advanced student to help another in difficulty has surely always existed. And yet tutoring is ancient! And we can draw up a long list of all the innovations, or things described as such, related to pedagogic situations, to show that, in matters of learning situations, almost everything has been done: individual tutors, group work, varied levels, needs-based groups, level-based groups, educational projects, written summaries, etc. In other words, the new is not intrinsically new, but what gives it the quality of being new is its situation in the context where it appears. For example, a school which has never before gotten its students to work in small groups announces that it is innovating, when this pedagogic technique is very old! We can take this innovation into consideration, although the practice is not, in light of the history of pedagogy, a novelty in itself!

We sometimes assign for reading a text on new education, describing an active pedagogy such as currently advocated, whose fundamentals are based on the ideas of Rabelais or Montaigne, or even Confucius, stipulating that "the child is not a vessel to be filled, but a fire to be lit" …

We hide the name of the authors and ask our students at what date such a text could be written. The answer is immediately agreed: the text is thought to be contemporary, which shows how, before innovating in pedagogy or at school, it would be useful to re-read the pedagogic classics, but that would go against the innovation which first of all invites us to reject the past[31]! The "new" of innovation in schools is thus always relative to the present and immediate context.

31 See section 1.1.2 "Link between innovation and tradition".

Thus, if a teacher has never had their students work on projects, they might declare themselves an innovator, even if there have been many precursors. This is the limit to which we can say that anything is new in schools, particularly when it comes to learning situations. It would even be possible to imagine a teacher declaring themselves to be innovative when nothing visible differentiates their pedagogy before and after, solely because they have changed their conception of their students. Defining the new is thus a tricky matter, and it is for that reason that no rules exist in the area such as we might find in organizational or structural innovations. We might even push the reasoning further and consider as an innovation a contemporary teacher giving corporal punishment to students, to the extent that this is a new practice and does not exist in the context surrounding such practices! And if we stretch the reasoning, the teacher might justify this innovation by the fact that the students seem to have learned their lessons better (evaluation). As to which deontologies and ethical purposes are at work: this will be the penultimate component of the definition of innovation, that of values, as we will see later.

Similarly, in the context of professional training of teachers, the last 20 years have been marked by the injunction to identify as a "reflexive practitioner", which consists of asking the teacher to reflexively examine their actual practices, to analyze them verbally in an adequate support process and to be able to identify the gaps to remedy them in a conscious and reasoned manner, even to innovate. Now, if we read the Spiritual Exercises of Ignatius Loyola from the 16th Century, we encounter this same type of practice which he calls "examination of conscience", to prepare oneself for an exercise of collective or individual discernment. Where is the newness, if not in the context and faced with more advanced knowledge in the field of psychology or psychoanalysis? For Husserl, value becomes the act of constitution of the object in a process of differentiation: this creates value and meaning. It is always value in relation to implied others.

1) So, you say to me, there is still something tangible that is new in schools, a new product or object, particularly information technologies, that which we call the introduction of the digital. These have been mentioned for at least the last 30 years as a means of revolutionizing teaching practices and learning. They are relatively recent and may qualify as "new", having in no way existed up until now in schools. Yes, but it is not the object in itself which makes something new, but rather the situation in which this object is used pedagogically. It is completely possible to use a computer or a tablet as

you would an image from a school textbook. These tools must therefore be on one hand "didacticized", and on the other, put "into pedagogical space". These demands relate to the competences of the teachers themselves. Innovation is thus found to be at the heart of the teacher–student relationship, in a pedagogical space conceived and designed by the teacher. Greater sophistication of digital tools will hardly replace a teacher's educational intuition.

2) The importance of goals must always be remembered in educational matters. The central question is to know what we wish for in a citizen of the future, with regard to those being educated. The current catchphrase is to "create society" or to "live together". However, what do these terms entail, to what pedagogical and educational conceptions do they correspond? Innovation in education most often arises from a feeling of dissatisfaction on the part of an educator or a team of teachers. This is increased by objectives or goals which are often ideals. We would say that, if the base of education is creativity, its essential motor, which sustains the energy of the innovator, is values. There is no innovation, especially in the field of education, without values which are sometimes intensified. This may correspond to wanting the best for another despite themselves, and this danger may befall some militants blinded by their good faith and the belief in their own rightness. Innovators often speak of improvement, of making things better; still, it is necessary to know what they mean by "better"! The road to hell is sometimes paved with good intentions, and it is without doubt that the values of innovation are embedded there, in this improvement. For example, to say: "I want to improve the wellbeing of my pupils" is so vague that good intentions are not enough; the aspiration for a better world sometimes leads to extremes worse than inaction!

3) Education is full of values and innovation and, when performed, is translated into concrete enactments of values rarely displayed by the institution. For example, desiring equity between students may indicate a perspective of giving more to those who have less, as has been done in France in the priority education areas (although this is a criterion linked to a residential area exhibiting social problems) or a perspective of dealing with differences among a heterogeneous group of students through external interventions, as also practiced in France by RASED[32]. All innovation in

32 The specialized teachers and national educational psychologists of RASED (French acronym for "Networks of specialized assistance for students in difficulty") provide specialized assistance to pupils in kindergartens and elementary schools in serious difficulty.

education pursues objectives and is underpinned by values. The problem is to analyze how these values are embodied in the innovative activities in question[33], and it is doubtless from here that most disagreements arise concerning school.

4) The fourth component relates to change. In reality, all innovation provokes change, but life itself is made up of change: we grow older, our hair turns gray, children grow up. In short, change is consubstantial with the passing of time and with life. Therefore, to say that innovation is change is a banality. On the other hand, we can say that innovation, if it is incontestably a change, is a particular change which might not have taken place without the desire of those who initiated it. In other words, innovation is voluntary, intentional and deliberate change. Now, some innovations come from outside: let us say that they are provoked by external constraints such as official educational reforms, but their innovative character comes from the way in which the teacher integrates or exploits them through their creativity and their will to change.

Innovation, based on creativity and on values, is based on a momentum towards the unknown, on an attempt to put into place an unscheduled activity, which appears to go in the direction of the desire of its initiators, to respond to a lack or a perception of deficiency. Innovation is not the only way to respond to a deficiency. Two others are a project, and problem-solving. Moreover, these two approaches are often assimilated to innovation, as they satisfy institutional demands for visibility of what the innovator wants to do, so as to help them financially or with support. These approaches are also forms of change, just like experimentation.

In fact, a project[34] is defined as a mental activity resulting in action governed by a certain number of realistic, often linear steps, with goals clearly marked out, a program, a timetable and a means of evaluation.

This assistance may be educational or rehabilitative. Their specific work, complementary to that of classroom teachers, allows a team to create a better response to difficulties in learning and in adaptation to the requirements of schooling experienced by certain pupils (source: education.gouv.fr, 17 March 2017).

33 We will see in the chapter on the evaluation of innovations in education that these goals are seldom achieved, and that other objectives that were not present at the beginning of the innovation appear and are achieved.

34 See the work of BOUTINET J.P., *L'anthropologie du projet*, PUF, Paris, 1990.

The approach of problem-solving is, for some, equivalent to that of innovation. Problem-solving consists of the identification of a perceived problem and searching for solutions to trial and evaluate their ability to resolve the problem. Innovation is of a different nature: it can exist without its author encountering any problem, simply because they wish to update their pedagogy, to try other methods of learning without yet perceiving any particular professional problem to be resolved.

One other approach can be added to these: that of experimentation, which is based on an experimental protocol revolving around variables to be monitored and evaluated through a controlled experiment. This wish to name innovation as something other than itself is a way to discipline it and to bring it into more identifiable frameworks of rational linearity.

5) Now, innovation is risk-taking, an adventure whose outcome we do not know in advance, and which surprises even the people involved. It is not reassuring for institutional officials, as they are obliged to accept this kind of maximum uncertainty. In other words, innovation is a process which is discovered as it progresses to completion. As the poet Antonio Machado said: "*el camino se hace caminando*"[35]. It is thus far from being a project, and using the expression "innovative project" somewhat distorts this adventurous aspect.

The advantage of innovation is that it allows the emergence of pedagogical activities unprecedented in their context (even if the new is somewhat diminished) and, in this case, context is the base of innovation. In fact, as mentioned previously, innovation is neither invention nor discovery, but the socialization of a new activity, that is, an appropriation by actors of this innovative activity (parents, students, teachers at the school, etc.). Innovation cannot institutionally develop except in a climate of autonomy, openness or even sometimes disorder; but we will examine this more closely in the chapter on educational policies and how they deal with school innovation.

In other words, a school innovation is an activity that includes something new in relation to its context, based on change which is voluntary, deliberate and intentional. It is a process steeped in uncertainty, based on creativity.

35 The path is made by walking it.

**Figure 1.6.** *Invariant features of the definition of innovation in schools*

However, a question still remains: where does innovation begin and end in the social activities and, in particular, the professional work of the teacher? No action is ever the repetition of another action; in every activity there are variations. We never exactly repeat the same thing, and the ingenuity of the teacher is quite often at work trying to deal with the unexpected behavior of students.

We remember one initial professional training for teachers which involved creating a teaching plan for 8th grade students. Each teacher–trainee thus had to carry out this project in the tiniest detail by specifying on a timetable what would be accomplished by the teacher and predicting the behavior of the students. Then, the enactment of this course project took place in conformity with what was planned, up until a student stood up impulsively and told the teacher that she was "pissing her off" with all these phrases!

This event had been in no way foreseen in the course project. What should the trainee have done? And it is here that the reactive capacity of the teacher intervenes, at the moment where nothing is planned and where urgent action is required to mobilize their inventive capacities, while still anticipating future consequences. And yet, it cannot be a matter of innovation, but of professional development to the extent to which the professional competences of the teacher are put into action in response to the field in which they work on a daily basis. These types of responses fall within the framework of competences of mastery. Thus, an expert teacher will know how to respond in the most appropriate way to these unexpected situations; the novice will develop these competences to the extent to which they practice their profession. The line between expert competence and the implementation of innovation is often quite fine.

Innovation begins with the voluntary, intentional and deliberate change, which we have discussed. Yet we must be careful, as one might consider any action that has not previously been performed, which is sufficiently isolated and recognized as new, to qualify as innovation.

And that might take place in the most mundane daily situations:

> "We innovate every time that we try something new, when we smoke our first cigarette, when we give a course for the first time without having bothered to prepare or when we discover that we can cheat on the test and not be found out. Of course, ordinarily, when we use an expression like innovation in teaching, we have the feeling of talking about something which is both new and beneficial. And we may even tend to want to claim that something is good because it is new or different from what we have done before."[36]

This understanding of innovation in relation to teaching can muddy the waters and give the impression that every novelty is an innovation. Now, the invariant features discussed above allow us to limit our frame of reference, even when all the invariants are not obviously present. For example, we are not persuaded that the process is an emergence, as the socialization of the practice of the action of smoking is strong and habitual.

36 HAVELOCK R.G., "Innovations in Education, strategies and tactics", Working papers, *British Journal of Educational Technology*, Ann Arbor, Michigan, 1971.

Other words surround innovation such as renovation, reform, adaptation, revolution, or refoundation. Let us take these concepts one after the other:

– a renovation is an activity consisting of restoring what exists (*re-novare*: to renew its attractiveness), in the sense of renovating a house. The latter remains the same, but it has a better appearance corresponding to its original appearance. It reinforces the character of the house without changing its meaning or usage. It is the same in education. A renovation is a change in organization and/or in pedagogical practice so that the goals, while being intangible, may be better achieved or improved in quality;

– a reform is a change carried out within the existing institutional framework (for example, agrarian or constitutional reform). Reform in training systems comes from laws and decrees, doubtless inspired by concrete observations, but the decisions are central, governmental ("top down") and aimed at allowing everyone to be reinscribed in a legal framework. The French educational system has seen an avalanche of reforms, quite often not allowing teachers to adapt to them, as another reform arrives to replace the previous. This is without doubt one of the causes of the inertia of French schools; too many reforms kill reform;

– an adaptation[37] is a response to a perceived change in the context. For example, students are not the same as those of 20 years ago, and teachers see themselves obliged to respond to the new defining characteristics of current students. They thus develop a professional practice in response to new conditions of existence as they perceive them; they adjust their activity to the circumstances which they encounter. The systemic approach is based on phenomena of regulation close to adaptation. Let us add that a reform may take the form of an adaptation;

– a revolution, etymologically, is an orbital movement around a central axis which brings the object in question periodically to the same point. Here, a revolution is seen instead as a radical upheaval, important and abrupt. It might even entail a violent transformation. Innovation is the opposite: it is the action of introducing something new in the intention of socially embedding it over the long term;

37 We are not talking here of adaptation as defined by Piaget, a phenotypic adaptation, which is "a process by which an organism transforms itself under the pressure of changes in the environment, or the result of this process, without this transformation being predetermined at the level of the organism's genetic system".

– a refoundation is a term more recently used in education to affirm the fundamental principles to be defended within education. It means to declare new foundations based on the original intention of certain policy directions. The word allows us to affirm a certain number of values attributed to education as priorities – namely, equality and solidarity. The affirmation of these principles leads to sets of actions based on four pillars, in relation to the law of 8 July 2013, published by the Socialist government of François Hollande, namely:

- educational success for all;
- a just and effective education system;
- students at the heart of the refoundation;
- trained and recognized staff.

From here also arise orders and decrees of application, which lead to the process of reform discussed above.

The abundance of vocabulary produced by the world of education shows, if it were needed to be shown, that the situation is complex and that moving the "mammoth" – as the former Minister of Education Claude Allègre puts it – is among the most complex and delicate of actions. The use of new words will not change anything while the educational system depends on wider political decisions linked to what the nation wants to be done with its schools.

## 1.4. Innovation as progress

In March 2017, a meeting took place moderated by Pierre Haski of the weekly *L'Observateur* with the suggestive interrogative title: "Is progress still progressive?"[38] The term "progress" is no longer as widely used as in the time of the Enlightenment and the last two centuries of history. It seems to have disappeared from the discourse of innovators as well of that of politicians.

38 See bibliography.

Peter Wagner, a German sociologist, was invited to this meeting after the publication of his book[39]. He declared:

> "We lived for two centuries with a certain conception of progress which no longer exists. We must rethink, reinvent progress, we still need it, but we can no longer return to the previous conception."[40]

This current pessimistic vision of progress is shown in the speech given by Albert Camus at the reception of his Nobel Prize, in 1957, when he said:

> "Every generation doubtless believes itself destined to remake the world. Mine, however, knows that it will not do so. But its task is perhaps greater. It consists of preventing the world from unmaking itself."

Now innovation, wedded to the search for something better as a horizon, is based on the idea that students, after having experienced innovation, will progress in their behavior and their learning. And reflecting on this, the word "progress" is singularly absent from current official discourse, while being frequently used by teachers; a student who progresses is a student who has better results. This idea of the linearity of progress remains; the student can only make progress when they are firmly in the grasp of the intentionality of the innovator.

If scientific and technological progress is seen positively through the material benefits brought to individuals (such as, we live longer, we can travel further and more safely, etc.), it is often thought of as engendering social progress which, however, leaves many left behind and produces more and greater inequalities. Social progress does not therefore seem to be on the agenda, and is linked to two parameters: domination and liberty.

It is interesting to note that the arrival of digital technology in schools has produced more hope than all the pedagogical practices mentioned in the preceding pages. This is due to the fact that all technology arrives with a

39 WAGNER P., *Sauver le progrès. Comment rendre l'avenir à nouveau désirable ?*, La Découverte, Paris, 2016.

40 "Le progrès est-il d'actualité ? ", *L'Obs*, no. 2731, p. 79, 9 March 2017.

strong promise of progress, even if disillusionment sometimes lies ahead. This is without doubt linked to the meaning given to its project, and to the need to believe that something is changing for the better.

Innovation is thus one of the ways of expressing this meaning to avoid having to accept the present era as it is, and to see it rather as "a reservoir of possibilities". The discourse of disintegration of the school seems to be nothing but refusing to create a new concept of progress. Innovation would thus be the expression of this collective intentionality (taken up by a collective or even by the whole establishment and beyond at its emergence), what Wagner calls "a collective democratic act".

## 1.5. Techno-economic innovation and innovation in schools: Tocqueville and Schumpeter

Innovation in schools is a concept which emerged well after innovation in technology and the economy. In fact, social evolution is primarily based on consumption and the accumulation of material goods. This is a major characteristic of capitalist societies, and it is possible to say that the cradle of development of innovation is the economic and industrial development of our societies. Profit is its essential motor.

Two authors, through their teaching, particularly marked this trend: Tocqueville with his conception of democracy as the future horizon of society, in particular America's, and Schumpeter, the precursor of the unveiling of an economic theory based on innovation.

Before becoming a positive doctrine, the demand for economic freedom was expressed as early as the mid-18th Century. This was the translation of a social aspiration: the desire for abolition of all the professional privileges which regulated access to the trades. This aspiration was in turn a response to the pressure of technical innovation and progress. In fact, economists such as Turgot or Mirabeau have shown that the corporative framework was a considerable brake on progress, as every appearance of a new product got a hostile reception from those professions whose reserved domain was threatened by it. For the citizen, this double corporatism of the monopoly of production constituted one of the factors in the cost of living: the abolition of privileges and the freedom of industry were the same thing in their eyes.

The individual thus acquired social power in a specific relationship with the law. It was no longer God, Nature or some other transcendence which determined the laws, but subjects relying on their will and their reason.

Tocqueville is one of the authors who speaks with the most clarity on this inexorable democratic movement, defined by "the decomposition of the aristocratic hierarchy of the past and the progressive equalization of the conditions of individuals"[41]. He bases his arguments on the American society of his time (the first half of the 19th Century), which he rates as the best democratic model based on the very fact that there is no hereditary aristocracy. Moreover, it is in turn, he says, because individuals have become equal before the law that their main activity has become economic: this is the reflection of the primacy of the economy that will remain the case all through the development of innovation. The latter becomes necessary as a source of adaptation for a capitalist society firmly anchored to the principles of individualism and the free market.

Building upon this later is Schumpeter who, after the crisis of 1929, offered his theory to the United States, finding in that country (a light-weight State and a high level of decentralization) the material for developing purely capitalist ideas. Thus, Schumpeter bases his theory on three interrelated organic poles: economic growth, innovation and technological progress. Innovation is central for him, as it constitutes the motor of growth and provides profits to firms to invest and produce still more and better. We are in the entrepreneurial world of technological innovations, generators of new products or new lines of production which are more efficient and more attractive to the consumer. The individual is seen as a consumer. When a product arrives at maturity, growth slows down. The latter is driven by a cascade of innovations which keep diverse businesses in the same market competitively fit. Each firm owes its survival and expansion to the development of innovation[42]. It lives from innovation to innovation, on the

41 ARON R., *Dix-huit leçons sur la société industrielle*, Gallimard, Paris, 1962.

42 This is shown in GARREL G. and MOCK E., *La fabrique de l'innovation* (2012), where the survival of the Swiss watchmaking industry depended on an innovation adopted just in time to avoid the layoffs of many workers, through the emergence of the Swatch watch whose process of innovation is narrated from its creation to its commercialization.

path marked out by technological progress, which itself is considered as the road to economic and social progress:

> "In fact, the fundamental impulse which puts the capitalist machine in motion and keeps it going is given by new consumer objects, new methods of production and transport, new markets, new types of material organization, all the elements created by capitalist initiative."[43]

And Schumpeter insists on this idea, finding all sorts of virtues in these innovations.

In the end, innovations are like exogenous variables which arrive in "clusters". Each innovation represents a cyclical process of installation with phases of expansion, recession and depression. The appearance of innovation ensures economic expansion and a temporary monopoly for the dynamic firm.

Schumpeter distinguishes five types of innovation:

1) the creation of a new good;

2) the introduction of a new method of production;

3) opening up a new market;

4) the conquest of a new source of raw material or of semi-finished products;

5) the realization of a new form of organization such as the creation of a monopoly situation.

It is out of the question, in his reasoning, for innovation to arise outside the economic and commercial domain; it is the very heart of his definition. Moreover, in the continuity of his ideas, he attributes a glorious role to the entrepreneur, this hero of modern times to whom he ascribes a wholly laudatory psychological profile:

43 SCHUMPETER J., *Capitalism, Socialism and Democracy*, Payot, Paris, p. 121, 1965.

> "For success, everything depends on the glance, the capacity of seeing things in a manner which experience subsequently confirms [...] finally the faculty of acting on others, which we may call by the name of authority, or weight. He creates without respite, because he can do nothing else [...], he possesses first of all the dream and the will to create a private kingdom, most often, but not always, a dynasty as well [...]. Then comes the will of the conqueror, finally the joy of creating a new economic form."

A short time ago, we still heard this discourse about business leaders and today to justify the massive sums paid in remuneration to CEOs of major public companies!

It is possible to identify five characteristics of Schumpeterian innovation:

1) it is endogenous to the economy in an exogenous process, and therefore it takes place as a phenomenon of economic and social development;

2) it arises independently of social relations which are only discussed as they relate to the innovation (opponents, for example);

3) it needs people who know how to promote its emergence and development (entrepreneurs);

4) it needs an institutional framework (justice, administration, etc.), which promotes the desire for innovation and science for economic purposes;

5) finally, it is beneficial for the economy and thus for society.

Knowing that capitalism is inseparable from private property, it is not certain whether, even for innovation in schools, such a paradigm is not currently being followed by some politicians, through the idea that schools could take part in the process of a competitive market, leading to a multiplication of innovations whose goal would be to improve the "student-product". The parents taken hostage thus become consumers and make a market choice between schools.

In fact, if education is the responsibility of the public service, it is not a sovereign function of the State like the police, justice, the army or printing money, because it does not relate to the security of goods or persons. In other words, as a public service, education could be provided by a private firm (conceived in the same way as that of Schumpeter) which would work, naturally, in the general interest under the control of the State, which would ensure respect for continuity of service, equality of treatment of users and other fundamental principles of public service. This would allow some room for liberty, even for abuses…

Such an idea poses the question of guaranteeing these principles and of how it would be verified. If the tendency of a country is towards privatization or deregulation of this public service, it is conceivable that we would witness the emergence of many so-called private, competing schools, following the spiral of innovations which, under the positive image of the development of young people, in fact only have one goal: that of profitability. Education, and in particular, the school system, is complex and vulnerable, so how can we envisage a deployment of innovations which does not follow this paradigm?

Doubtless here we can affirm that innovation in schools cannot be reduced to techno-economic innovation.

There did not yet exist in Schumpeter's time the vast and formidable field of advertising, which often creates artificial needs among consumers.

## 1.6. Innovation and its process of socialization

Let us recall that innovation follows invention; we might say that innovation is an invention that succeeds, for not every invention results in an innovation. It may happen that an invention remains on the shelf, or that it emerges because the social context is ready to accept it. This is, for example, the case for the electric toothbrush, invented in 1939 and only commercialized (as the Broxodent) in the 1960s. We might also mention what Akrich[44] discusses, that is the appearance of frozen food in 1912, whose use was not socialized until around 1952, in other words 40 years later!

In fact, the central feature of innovation is found in this process of socialization, that is, in its adoption by society. The field must be able to

44 AKRICH M., CALLON M., LATOUR B., "A quoi tient le succès des innovations ? 1) L'art de l'intéressement 2) Le choix des porte-paroles", *Gérer et comprendre*, Les Annales des Mines, pp. 14–29, 1988.

accept the use of the new product which will then spread. The social process is eminently important; it may be based on the tides of fashion (for example, in schools, non-directive pedagogy based on certain libertarian tendencies and on a misunderstood Rousseauism, fashions which might subsequently be violently rejected and accused of all manner of social crimes such as the loss of authority of teaching staff, absence of discipline in classrooms, etc.).

This process of socialization is uncertain and may flare up like a gunpowder trail, disappearing with as much rejection as it was adopted with excitement. This is doubtless one of the defining elements of innovation in schools: what was cherished several years previously is often reviled.

Moreover, social practices and the use of new objects are part of a social imaginary, that is, a dynamic system of facts and images, of social representations and grand narratives. This system works by its own, always ambivalent logic. It fulfills the function of linking concept and percept, and enriching reflection and perception. This imaginary, or techno-imaginary when it comes to new objects, particularly digital objects, is formidable and uncertain, because it does not follow market logic.

It is therefore very difficult sometimes to tell the story of the birth and deployment of an innovation, without falling into Schumpeter's imaginary of the brilliant and heroic entrepreneur or the strategy of a foresighted politician or the great creativity of an engineer or a group of people. It is all of this at the same time, in disorder and contingency.

For example, Fordism or Taylorism are organizational innovations and, to be established, they had to defend their point of view against unions, against workers and against bosses. However, what often tips the scales towards the implantation of innovation is society in general, unconsciously searching for a solution which satisfies its expectations and agrees with its imaginary. Similarly, there have been many commercial innovations over the past 20 years such as hypermarkets, bulk-purchase stores and drive-through shopping. How to explain their success, of which we are now sensing the decline?

Two theories allow us to understand how they come about. The first is the mechanistic theory of Fisher,[45] which sees innovation as an exogenous

45 FILSER M., "Méthodologie d'élaboration d'une typologie de clientèle en marketing direct", *Revue Française du Marketing*, vol. 126, no. 1, pp. 57–66 1990.

process, showing that innovation is a response to an environmental demand, no matter the structure of the proposal, from the moment that this innovation comes to the point of implantation, in a process of adoption. Society is an important actor in the deployment of innovation, so much so that, in innovations linked to the commercial domain, this process is called organizational socialization of the client (OSC)[46], something close to a process of learning. We can also mention another theory, that of the forms of implication of social actors in this socialization of innovation, that is that the logics of actors are of primary importance in the development of innovation.

These two tendencies are different. We consider innovation through its intrinsic qualities (that is, indisputable performance) and the other, more sociological, considers innovation through its implantation, as the result of a compromise between its characteristics and the demands of users. Within the latter tendency, social influence and the role that might be played by minorities[47] discussed by S. Moscovici (more in the order of social psychology) and the contributions of Callon (more from a sociological approach) deal centrally with the movements of social groups and power relations.

In fact, the work of Akrich and his team[48] indicate that the future of innovation depends on the interest that the latter evokes among "a whole series of social groups who will decide on its future" (p. 20). The authors add:

> "The model of engagement underlines [...] the existence of a whole bundle of links which unite the object with all those who manipulate it. The model of diffusion locates the technical object within a society which constitutes a more or less receptive environment. The model of engagement introduces all the actors who welcome the object or turn away from it, and it emphasizes the points of intersection between the object and the more or less organized interests whose attention it attracts." (pp. 21–22)

46 GOUDARZI K., La socialisation organisationnelle du client dans les entreprises de services, PhD thesis, University of Aix-Marseille III, 2005.

47 MOSCOVICI S., *Psychologie des minorités actives*, PUF, Paris, 1991 (2nd edition).

48 AKRICH M., CALLON M., LATOUR B., "A quoi tient le succès des innovations ? 1) L'art de l'intéressement, 2) Le choix des porte-paroles. Gérer et comprendre", *Les Annales des Mines*, pp. 4–17 and 14–29, 1988.

In other words, the success of an innovation depends on a sort of aggregation of interests "and on the participation of all those who decide to promote it". This is what Callon and Latour call the entry into the "spaces of engagement" of individuals.

The socialization of innovation is thus linked to the will and ability of actors to adapt the innovation in their place of social activity: this is a strong mobilization, often through collective work. Akrich adds: "Adopting an innovation means adapting it" (p. 1).

It will thus be necessary first to identify the bearers of an innovation, the leader-clients, in other words: "the users who are best placed to transform the innovation and to bring it towards what users will demand" (p. 3). The latter will thus be responsible for creating a favorable context for the diffusion and acceptance of the innovation. It is up to the organization to adapt its innovation and/or convince these users. And the authors stress the need to bring about a co-evolution of the object and society, that is, the technology and its public, in a "work of mutual adaptation which determines adoption" (p. 6).

The joint work of this engaged collectivity and the organization will imply adaptation of the innovation through "all-out experimentation through successive iterations" (p. 6), which is to say, a process of back-and-forth between the actors who negotiate the characteristics that the innovation must possess. Thus, "the innovation is transformed at every iteration, redefining its properties and its public" (p. 8), so as to take account of its publics' expectations, but also of the evolutions of the potential market. However, this assumes that the original public of the innovation, those who are involved in these negotiations, are good communicators, that they are relevant and legitimate. Even then, it involves a crucial decision whose relevance will not be perceived until the end of the process. Now, this choice will lead to other choices such as "strategic directions, but this also involves choosing what we will innovate on and the problems that will need to be solved" (p. 15).

Now for Akrich *et al.* (1988) through this process, it is a question of "getting into a position to formulate relevant questions, that is to translate an economic strategy into research actions [... but also] to find explanations and discover rules for action" (pp. 17–18).

Rogers' (1995) theory of diffusion of innovations is somewhat different: the author seeks to identify how an innovation spreads. Through the example of a failure (the diffusion of the habit of boiling water in a Peruvian village[49]), Rogers stresses that the diffusion of an innovation is favored by the compatibility of the innovation with the values, the beliefs and the lived experiences of the individuals who compose the social system in which the innovation must become established. In fact, the author underlines "the importance of interpersonal networks for the adoption and rejection of an innovation [to the extent that] diffusion of innovation is a social process, as well as a technical issue" (p. 4). Now, this process affects the way in which potential adopters will perceive the bearer of this change. Consequently, this process also affects the willingness of potential adopters of new ideas. Based on this example, the author defines diffusion as "the process by which an innovation is communicated through certain channels over time between members of a social system" (p. 5) and communication as "a process through which participants create and share information with each other in the goal of arriving at mutual understanding" (p. 5).

From this he deduces that diffusion is a matter of communication, in the sense that innovation is contained in the message. Now the particularity of this message resides, according to him, in the uncertainty of its diffusion among the population in question. Consequently, the diffusion of innovation can be understood as "a sort of social change, defined as the process by which an alteration occurs in the structure and function of a social system" (p. 6). This diffusion may be spontaneous as well as planned. Relying on this idea and these definitions, Rogers (1995) offers a definition of each of the characteristics of the diffusion of an innovation: the innovation itself, communication through certain communication channels including advertising alongside virtual social networks, time and the social system.

49 The failure referred to took place in the Peruvian village of Los Molinos in the coastal region. Illnesses due to water pollution were increasing, from which arose the idea of implanting an innovation: that of boiling drinking water. However, at the end of a two-year intensive campaign, only 11 homes out of 200 were boiling their drinking water. Two factors explain this failure: on the one hand, the level of knowledge of the population, who did not know what germs were and who saw water as clean and if there were small creatures in it, they would be like ants, not dangerous to humans; on the other hand, the relationship with boiled water was determined by dietary regimes where local beliefs attributed a particular temperature to foodstuffs, independent of their actual temperature.

Rogers (1995) defines an innovation as an idea, a practice or an object, perceived as new by an individual or a group of individuals who might adopt it. According to him, it is the perception of the novelty of an idea from an individual's point of view which determines their reaction to it – that is, its adoption or rejection. Consequently, if an idea is perceived as new by an individual, that is, if it lends itself to understanding, persuasion or a decision to adopt, this will be considered as an innovation.

Moreover, according to Rogers, the process of decision-making with regard to an innovation essentially resembles a process of research and information processing, through which the individual is motivated by the reduction of uncertainty as to the advantages and disadvantages inherent to the innovation to be adopted. Thus, based on the information available concerning the following characteristics of the innovation, it will be more or less rapidly adopted:

– Relative advantage: to what extent is the innovation perceived as better than the idea which it aims to replace?

– Compatibility: to what extent is the innovation perceived as compatible with the existing values, past experiences and needs of the individuals affected by the adoption of the innovation?

– Complexity: to what extent is the innovation perceived as being difficult to understand and use?

– Is it possible to trial an innovation: to what extent is it possible to experiment with the innovation on a limited basis?

– Are the results of the innovation observable: to what extent will the innovation be visible to others?

Furthermore, independently of these characteristics, the author indicates that an innovation may evolve over the process of its diffusion, thus echoing the argument of Akrich *et al.* (1988) in the sense that innovation will in some way adapt itself to its public in order to promote its adoption.

Rogers (1995) indicates that communication is doubtless one of the most important factors in the process of diffusion of an innovation. According to him, the very essence of the diffusion process lies in the exchange of information, through which individuals transmit a new idea to one or more people. Now, depending on the nature of the exchange, its impact will not be the same. Thus, media allows an important audience to be reached, while interpersonal exchanges are more effective for persuasion regarding the quality of an idea. The effectiveness of interpersonal exchanges will be all the stronger if the exchange takes place between individuals of similar economic status and education.

Relying on studies of the diffusion of innovation, Rogers (1995) stresses that early adopters are sensitive to objective evaluations of the innovation, while later adopters are mainly influenced by subjective assessments transmitted by their peers who have already adopted the innovation. This finding suggests that the process of diffusion resembles the imitation of early adopters by later adopters. Consequently, diffusion of an innovation depends on the experiences of the first adopters.

Rogers (1995) stresses that one of the essential problems of diffusion of an innovation is that communication of messages about an innovation generally takes place between heterogeneous groups of people, which means that this communication is inefficient. He considers time to be an important element in the diffusion of an innovation. According to him, the process of an individual deciding on the adoption or rejection of an innovation can be conceptualized through five successive stages.

First, the group of individuals must become aware of the existence of the innovation, to understand its functions. At this stage, mass media may transmit this type of information. Next, the group will take a positive or a negative attitude towards the innovation. The resulting attitude will consequently imply the adoption or rejection of the innovation. This attitude is based on the evaluation of the innovation through interpersonal networks. Depending on the preceding stage, the individual will put the innovation in question into use, or not.

Thus, following the decision on adoption and testing the innovation, the individual will either confirm their decision or not – that is, the test will lead either to reinforcement or reversal of the decision taken previously. They may also decide to adopt the innovation and then to reject it (*discontinuance*[50]), whether because they were not satisfied by it, or because it has been replaced by a better idea. They may also adopt it at a later stage.

50 We deliberately employ this term which gives a more precise meaning of suspension of a continuity.

Rogers classifies the different categories of adopters of an innovation on the basis of the time required for this adoption into five types: innovators, early adopters, the early majority, the later majority and the laggards, or recalcitrants.

The curves of Figure 1.7, as presented by Rogers, illustrate this process of development of innovation.

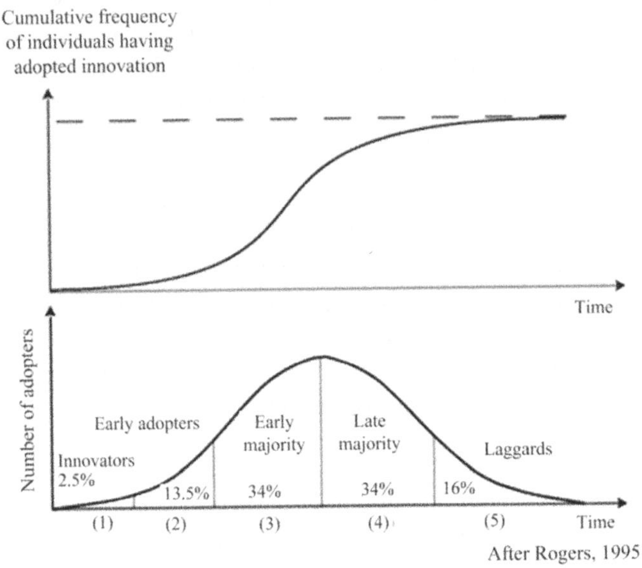


**Figure 1.7.** *Curves of the process of development of innovation according to Rogers*

Innovators are active researchers of information relating to new ideas. They have a high degree of access to media and their interpersonal networks are very extensive. In addition, they are capable of facing a higher level of uncertainty than the other categories with regard to an innovation. They are even more important in that they are the first to adopt an innovation: they cannot therefore rely on the subjective assessments of their peers. More widely, innovators and early adopters (generally young and educated, but not always!) tend to influence members of the other groups. In this sense, they may be considered as opinion leaders for certain categories of products (here we can see the "spokespersons" mentioned by Akrich, Callon and Latour (1988)).

Finally, the members of the other groups – that is, the later majority and the laggards – tend to be more cautious in buying new products and tend to wait longer to adopt the latest innovations. Rogers argues that most innovations follow an S-curve. In fact, if we place the cumulative number of individuals adopting a new idea on the Y-axis and time on the X-axis, the resulting distribution forms an S-curve, as in Figure 1.7.

In other words, in the first period, only innovators will quickly adopt an innovation. Then, the curve begins to rise: this corresponds to the adoption of the innovation by more and more individuals. After a certain stage, the trajectory will stabilize, because fewer and fewer individuals will remain who have not yet adopted the innovation. Finally, the S-curve reaches its asymptote, the process of diffusion ends, and we have reached the institutionalization and the banalization of the innovation, that is, its disappearance.

Rogers shows that the diffusion of an innovation is affected by the social system, in the sense that the latter may pose a barrier. It is therefore necessary to consider the effects of the innovation on existing norms, the role of opinion leaders and the consequences of the innovation. In fact, the structure of the social system may favor or prevent the diffusion of an innovation.

Thus, each person's propensity to adopt innovation is affected on the one hand by their individual characteristics and, on the other hand, by the nature of the social system surrounding the individual. The adoption of an innovation is also affected by the norms in force. The latter may be found at the level of a nation, of a religious community, of an organization or of a local system such as a village.

Consequently, the diffusion of an innovation will rely on opinion leaders. They illustrate and express the structure of the social system within which they exercise informal influence. This is explained by the fact that their opinion-leadership is gained and maintained by their technical competence, their social accessibility and their conformity to the norms of the social system in which they develop.

In some cases, the diffusion of innovation will be carried out by what Rogers (1962, 1995) calls a "change agent". This means a professional whose goal is to obtain the adoption of a new idea. This latter is generally an individual who will have to rely on opinion leaders. He will also rely on the latter to reinvent the innovation, that is, to carry out changes or modifications to the innovation through its process of adoption and implementation. In one sense, the opinion leader will partly perform the role of a translator of the innovation among their peers, in the meaning of Akrich *et al.* (1988). They can promote adoption by their peers by promoting its adaptation and by explaining it to them.

What stands out in our review of the literature is that innovation as it is commonly discussed – that is, in firms – owes its success to the entrepreneur and his creativity (Schumpeter, 1939). In addition, to succeed, the organization must translate and adapt this innovation to its intended public (Akrich *et al.*, 1988). In this context, Rogers (1995) stresses the importance of peers and of time for diffusion of information about the innovation as well as the steps of adoption. Finally, the author sorts information channels according to the public to whom they are addressed (innovators, early adopters, early majority, later majority and the laggards). These channels are even more important in that according to Rogers (1962, 1995), the process of decision-making concerning an innovation essentially resembles a research and information-gathering process.

We will now proceed to identify the factors which may favor the success of an innovation among consumers. This question leads us to focus on the *Technology Acceptance Model* (TAM), which identifies perceived usefulness and ease of use (Davis *et al.*, 1989; Davis, 1989) as constructs favoring the adoption of a new technology.

The perspectives of the authors discussed, if not diametrically opposed, seem to have different orientations and varied conceptions of society. For some, society is an entity that shapes itself based on the intervention of personalities, whether heroic (Schumpeter), or "change agents" well-placed at the intersection of the technical qualities of innovation and the composition of society (Rogers), or a very diverse society with groups of individuals who become involved in controversies and sit in judgment, and who must be involved by taking account of their space of engagement according to a complex strategy (Callon).

These three great conceptions are no doubt themselves in step with the evolution of society, and Schumpeter's concept of democracy, largely inspired by that of Tocqueville, is very different from our current democracy as shown in the works of Akrich *et al.*

Other, lesser authors have also discussed a conception of diffusion of innovation, such as Davis *et al.* (1989)[51] or Childers (2001)[52]. The former author speaks more about organizational innovations; he focuses on the receptive public and on the perceptions they may have of an innovation's degree of usefulness and its accessibility without too much effort. He names this process the TAM (*Technology Acceptance Model*) and summarizes it as in Figure 1.8.

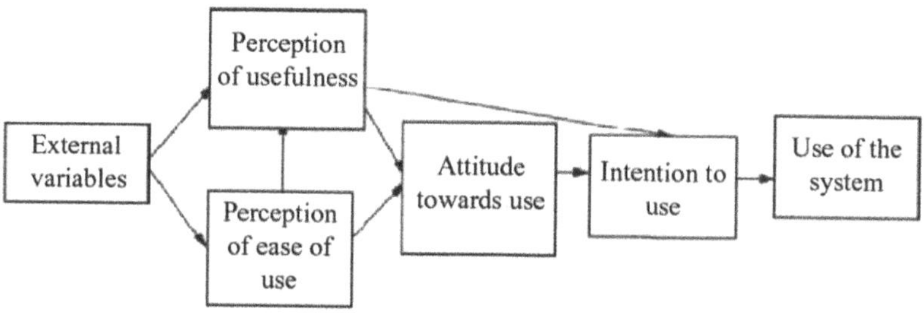


**Figure 1.8.** *Model of acceptance of technological innovation*

In other words, it seems that the literature devoted to the adoption of innovations concentrates on the transaction (Davis *et al.,* 1989; Davis, 1989; Childers *et al.*, 2002; Venkatesh *et al.*, 2003) and on the relationship, including the context in which the innovation[53] appears (Akrich *et al.*, 1998; Rogers, 1995) and then on concepts such as subjective norms which seem to play an important role (Ajzen and Fishbein, 1980)[54].

To these latter authors we can add researchers for whom the primary component of innovation appears to be subjectivity: fundamentally, it is human behavior, in the adoption of a new activity or a new technology,

51 DAVIS F.D., "Perceived usefulness, perceived ease of use, and user acceptance of information technology", *MIS Quarterly*, vol. 13, no. 3, pp. 319–340, doi:10.2307/249008, 1989.

52 CHILDERS T.L., CARR C.L., PECK J., CARSON S. "Hedonic and utilitarian motivations for online retail shopping behavior", *Journal of Retailing*, vol. 77, no. 4, pp. 511–535, 2001.

53 In fact, can we imagine the diffusion of the use of the spade in a society where individuals did not have shoes?

54 AJZEN I., FISHBEIN M., *Understanding Attitudes and Predicting Social Behavior*, Prentice-Hall, Englewood Cliffs, New Jersey, 1980.

which subjectively invests it with novelty, and the elements upon which this is based may be of the order of confidence (Gefen *et al.*, 2003)[55], of motivation or of emotion (Venkatesh, 2000)[56].

Opening the way for more experimental research, Benbasat and Barki (2007)[57] concern themselves with the patterns of use, reinvention and learning behaviors which users develop. The idea is that the TAM approach remains superficial and does not consider the possible diversion of an innovation by its users, and even the transformation of the initial goals of the innovation. In this same vein, it would be interesting to see in what we call "diversion of an innovation" perhaps a form of learning and of production of new knowledge given a new object. As Bernoux recalls[58]:

– in 1960, we worked for the product;

– in 1980, we worked for the client;

– in 2000, we worked with the client.

We can sum up these different approaches in Table 1.1.

Curiously, no theory of diffusion of innovation in schools has been produced, to the extent that we might raise the question of whether innovation in schools can be treated in its conceptualization as a technical innovation, a social innovation, an information systems innovation, or whether it is quite simply not applicable within an institution whose goal is the production among young people of social norms capable of maintaining a viable and cohesive society.

Would it be an oxymoron to talk of innovation for education, in the sense that education does not in any way resemble a firm with respect to its goals or function, nor a service-providing institution such as retirement homes, non-profit organizations or social assistance?

55 GEFEN D., KARAHANNA E., STRAUB DETMAR W., "Trust and TAM in Online Shopping: An Integrated Model", *MIS Quarterly*, vol. 27, no. 1, 2003.
56 VENKATECH *et al.*, "Age Differences in Technology Adoption decisions: Implications for a Changing Workforce", *Personnel Psychology*, vol. 53, pp. 375–403, 2000.
57 BENBASAT I., BARKI H., "Quo vadis TAM", *Journal de l'Association pour les systèmes d'information*, vol. 8, no. 4, pp. 211–218, 2007.
58 BERNOUX P., *Sociologie du changement*, Le Seuil, Paris, 2009 (new edition).

| Author | Origin of the innovation | Diffusion and deployment of the innovation |
|---|---|---|
| **Schumpeter** | Heroic entrepreneur | Entrepreneur's capacity to convince |
| **Rogers** | Innovator (individual or industrial group) | Epidemiology, contagion. Role of a change agent allowing its passage from pioneers to laggards |
| **Akrich, Callon and Latour** | A team of innovators which gather future users through the process | Continuous engagement including at the moment of origin of the innovation. Adaptation to the public |
| **Davis, Childers, Venkatech** | From the outset of the innovation, the user is envisaged in the virtual process of use as a transaction (TAM model) | The innovation includes in its presentation the usefulness and ease of access that it may have for users |
| **Ajzen, Gefen, Venkatesh** | Producers of innovation | Appropriation is linked to the affective investment (confidence, motivation) that can be made in the innovation |
| **Bensabat and Barki** | | Reinvention of an innovation by reappropriation, going as far as possible, diversion of the initial goals of the innovation |

**Table 1.1.** *Different approaches to innovation*

We are always at risk of saying that innovation in schools follows a process close (but not identical) to that of other social organizations such as health or social assistance. It doubtless develops through various spaces of engagement, motivation and perceptions of use, which allow us to understand its adoption.

Now, the reasons that educators might be motivated to innovate, to see it as useful in education, may be very numerous. The problem is that they are often based on contradictory, even opposing ideologies, about which the least that can be said is that trying to overcome them is like squaring the circle!

And the heterogeneous composition of the representativeness of the personnel engaged in teaching, through the different categories of actors involved in the education system, will not make this cohesive construction any easier, since their interests are often divergent.

The originality of the theory of engagement, in understanding the development of innovation such as it might occur in schools, comes from the fact that it is not an aggregation of various kinds of people (which corresponds to the composition of staff in education), which allows the socialization of an innovation, but the internal organization of their understanding in such a way that all of them find an interest in it.

The theory of Habermas in terms of communicative action might be useful in finding a way out of this cacophony[59]. Habermas, a researcher of the Frankfurt School, aims to get out of the dilemma of social cacophony which is most often resolved by injunction or social influence (which he calls strategic action). He proposes that the collective action of engaged people for a common result (here, the success of the students) should be based on communicative action, oriented towards a common understanding. He highlights reason as the source of mutual understanding, reason rooted in language and discourse. Thus, in the case of disagreement in the "lived world" (for him, the concrete professional environment of the individual), individuals must interact verbally, calmly, through a discourse which claims three things: a claim to precision, a claim to justice with regard to the social context and its norms, and a claim to sincerity. In other words, it is a matter of coordinating social interactions in advance and not only their effects. He calls this negotiation.

The role of the State is thus only to extend the innovation in such a way as to materially facilitate its general implantation. In fact, a State or a

59 HABERMAS J., *Théorie de l'agir communicationnel*, volume 1: *Rationalité de l'agir et rationalisation de la société*, J.-M. FERRY (trad.), Fayard, Paris, 1981; volume 2, *Critique de la raison fonctionnaliste*, J.-L.SCHEGEL (trad.), Fayard, Paris, 1987.

department cannot innovate, because that would imply taking ownership of this innovation and thus perverting it. It can only take ownership of the decisions that it makes. We will see this more clearly in the chapter on the links between educational policies and innovations at school.

## 1.7. Innovation and its future

The question raised here is to ask whether an innovation adopted after a certain length of time remains an innovation. What does an innovation become in the longer term?

Schumpeter talks of "creative destruction", that is, simultaneously with the emergence of innovation comes the disappearance due to obsolescence of sectors of economic activity jointly with the creation of new activities, all these based on two fundamental ideas: the idea of capitalism as the base of understanding, and that of technological progress.

In other words, an innovation is called upon to disappear in favor of another which allows technical and social progress. If innovation destroys, it is to produce better, more efficiently. This destruction may be violent, but this violence is softened by the arrival of a supposed improvement, that is through the new innovation. The question raised is that of how to judge what is better, of which we have already spoken in the definition of innovation: who judges "better" and who benefits from it, even if it proves to be unavoidable?

Several decades ago we witnessed an exponential explosion in the number of innovations, as shown in Figure 1.9.

The banalization of innovation, particularly in the social realm and doubtless in schools, leads us to think that this multiplication of innovations of different sizes leads to actors developing competences in innovation. In other words, innovation produces competences in the form of learning, producing in turn more innovations, in an ascending spiral. In any case, the competences acquired during this innovation would remain.

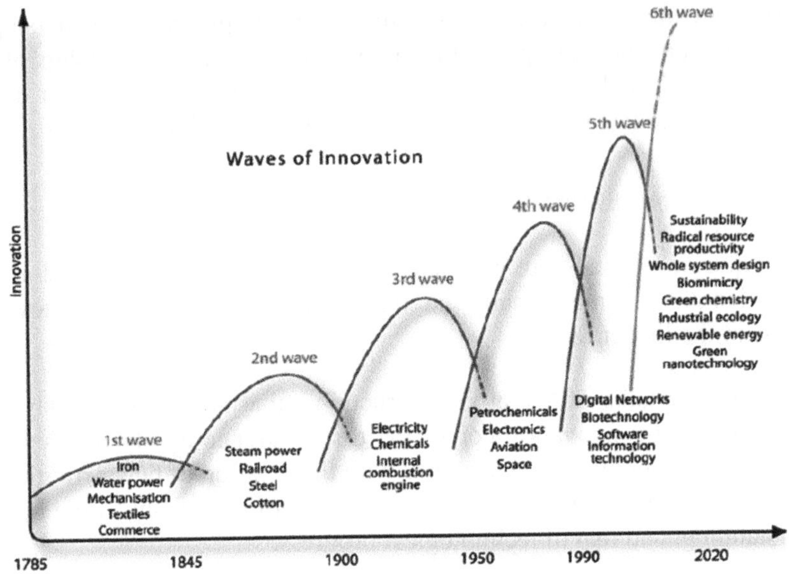


**Figure 1.9.** *Change in the number of material innovations from 1785 to 2020 (source: The Natural Edge Project (2004))*

Innovation may also disappear due to pure disinterest or be considered obsolete, or even dangerous. In schools, innovation may have three fates, namely:

1) It dies a quiet death due to lack of investors (for example, if the bearer of innovation is silenced or retires, or when the collective of innovators enters a phase of conflict). Some traces may remain to be taken up later by another innovator.

2) It becomes ossified, especially in the case of social innovations which tend to remain in a corner without anyone either being offended or taking up the torch of the innovation. We have thus known in schools some innovators who have been innovating for more than 20 years.

3) It becomes institutionalized, that is, it is taken up by the institutions who extend its boundaries, even making it compulsory. We will see later what kind of links form between innovation and institution for the generalization of an innovation. For example, the configuration of classes in schools, that is, the grouping together of young people of the same age for teaching at the same level, was originally an innovation which the

department seized upon and made the general structure and basis of the French educational system.

This last outcome is the most positive for the innovation if it wants both to become established and integrated into the social fabric.

The issue remains that an innovation is always adapted as it is adopted, that is that it loses its original "purity", often leading to the innovators refusing the paternity of the new innovation. Opposition to the development of the innovation may even arise among the original originators, who then struggle against its establishment and generalization, which might have seemed to be the goal!

## 1.8. From technical innovation to social innovation

### 1.8.1. *Different innovations: characteristics and implications*

The word "innovation" is often accompanied by an adjective which determines its boundaries, such as socioeconomic innovation, organizational innovation, pedagogical innovation, school innovation, technical innovation, technological innovation, social innovation and educational innovation. These adjectives define the field on which the innovation develops, that is, the field of socio-economy or of the organization, or of pedagogy, or more widely in the world of schooling, or in the domain of technical production, or that of technology which might be information and communication, or the field of social activity.

In addition, innovation might be a rupture, a process, incremental, ample, frugal, etc. It is a matter of designating the very quality of innovation. Incremental innovation signifies an innovation that involves a small degree of change and thus represents a slight disturbance in relation to what is usually done. For example, in a school, a teacher, who does not think they can teach an English accent to their students without voice work, invites his colleague who teaches music to intervene to help develop the students' voices. The objectives of the school are unchanged, it is only the way of implementing them, yet without disturbing the school establishment.

By contrast, an innovation which is a rupture is more radical; it changes the habits of life. For example, the introduction of professional-style multidisciplinary projects into vocational high schools leads several teachers to work together and to change roles, to become companions rather than providers of knowledge. This is both a change in role on the part of the teacher and a change in the very content of the disciplines. This innovation is very disturbing in its implementation.

As for frugal innovation which we will return to later, it invites us to change to do more with less. It is linked to a desire to combat waste (for example, recycling wastewater) or reduce energy consumption (turning off streetlights in a city after midnight). In schools, this frugal innovation has never really been considered. Is it, however, relevant to schools?

### 1.8.2. *Technical innovation and social innovation*

All over the world, mentioning innovation leads inevitably to the work of Schumpeter who doubtless shaped the destiny of innovation as a cumulative process taking place in a competitive economic space.

Moreover, the very evolution of the meaning of innovation has above all developed from the viewpoint of producing new consumer items. Research works discussing innovation often base their analysis on the production of a new product[60]. The idea of leaders or frontrunners of innovation is very current, even if these people do not always qualify as entrepreneurs in Schumpeter's terms. Start-ups are the most visible example today.

This production of a new object destined for consumers is a technical innovation in that the very basis of its existence is a new object which will be introduced onto the competitive market, derived from manufacturing production. The object comes first, and its social adaptations are relegated to the background, even if, as we have seen in section 1.6 on diffusion of innovation, the public is always present.

---

60 Even the works of sociologists, more focused on studying the innovative process and its impacts on the social conditions of individuals, mostly envisage innovation as production of a new object. This is the case in the research of the sociologist G. GAGLIO, particularly *Sociologie de l'innovation*, no. 3921, Que sais-je? PUF, Paris, 2014.

On this subject, we make a distinction between techno-economic innovation and social innovation. These two forms of innovation allow us to identify the priorities of innovation. In the first case, the focus is on the consumable object; in the second, on the provision of new services. Even if the economy is not excluded from the latter, it is not crucial to its beginning.

Techno-economic innovation transforms over time and the fact that two decisive dimensions are involved in production of the object – that is, the economic costs of operation (and forecast profits) and the discovery of consumer needs – moves the manufacturing aspect towards greater consideration of social factors.

The passage of time reveals complementarities between these two types of innovation. In fact, all technical or economic innovation requires the social aspect of implantation, and this is increasingly the case given the growing responsibility of civil society in economic and political decisions.

Although little valued over the years, social innovation is older than sociotechnical innovation. In fact, looking at the history of innovation, the concept has had a more political connotation (the direction of society) and the Greeks were the first to use the concept *(kainotomia)* to designate the introduction of change into an established order. However, this social innovation was regarded as disturbing and subject to manipulation, thus negative. The aspect of social governance has always been a complex issue, suspect in the eyes of citizens, and therefore delicate both to understand and to guide.

For example, we find in the *Encyclopedia Britannica* of 1888 (p. 211) that "communism is the name given to manipulation performed through social innovation whose central goal is to seriously threaten private property".

At present, social innovation is seen as a way "to develop new answers to social needs which are badly or not at all satisfied in the current conditions of the market and social policies"[61]. This author adds that "the voluntary sector is the historical laboratory of social innovation, it is its DNA".

61 P. GORGET, manager of the AVISE program.

Dandurand[62] identifies three phases of the evolution of the concept of social innovation:

1) limitation, which is the assimilation of social innovation to the concept involved by Schumpeter's approach, that is, a technical trend;

2) generalization, towards the end of the 1990s when we really witnessed a strong emergence of social innovation and its indissociability from technical innovation (seen as its corollary to the extent that traditions or behaviors would change with the introduction of use of a new object, where it would be necessary to combine existing habits with new disruptions);

3) the current phase of fragmentation of the concept of social innovation depending on the nature of its application. Moreover, it is here that the author situates pedagogical innovations as particular forms of social innovation.

Social innovation is not directly based on production of a new, competitive object, but it attempts to respond to social needs that are new or not currently being met by the market or by social policies, implying the participation and cooperation of all the people concerned. Social innovation leads to a new mode of governance of institutions, which poses a problem for the very structure of the French national education system.

The social power of social innovation is focused on the representations of actors and on their relations: in other words, building a more just society, more concerned with the development of the subject. Innovation in schools resembles this conception, as it defines itself at the heart of learning and in a particular social relationship, the pedagogical relationship.

In France, social innovations have most often taken the form of voluntary associations formed under the law of 1901, where several people combine to respond to a new, unsatisfied demand, strongly involving their target audience in their new proposals and ideas. These associations have, moreover, often been criticized for allowing public policies to continue to ignore these new needs, since these associations are taking responsibility for them. This is a wide debate on which we will not go into more detail. The judicial aspect of this law of 1901 allows these associations to operate democratically; the crucial problem is their funding, which is often

62 DANDURAND L., "Autour du concept d'innovation sociale, approche historique et comparative", *Revue française d'Administration publique*, vol. 3, no. 115, 2005.

indicative of their political and philosophical orientations. Some politicians[63] have even envisaged that social innovation should become the regulator of society, responding to its rapid evolution, even considering creating a fund to help their development. These associations are not funded by profits, requiring funds to be raised from members, from patrons and/or from State services.

Social innovation rules in those sectors where the State is not able to fully carry out its role, for different reasons. These sectors may be: youth unemployment, energy, housing, the environment, providing food or questions of over-indebtedness of households.

Education is also an area for this social innovation when it involves, for example, giving a "helping hand"[64] to children from disadvantaged backgrounds with their homework. In all these social innovations, the users themselves are the origin of innovation; they do not become the consumer of an innovation produced elsewhere, but the creator of innovation through dealing with the shortfalls they encounter.

In France, the social and charitable economy sector has long experience in this field. It saw the blossoming of several associations which have formed a national movement[65].

However, what really raises the question of innovation is *jugaad*[66] or frugal innovation which, in spirit, is doubtless not as new as some might suggest, but leads us to rethink innovation in both its form and its objectives.

### 1.8.3. *Innovation of withdrawal and frugal innovation*

Generally, innovation is seen as the arrival of a new element in an existing context. This conception is based on the idea that innovation

63 It is enough to recall the intervention of François Holland on 30 April 2013, where he declared that the State should financially support certain associations developing social innovations.

64 "Coup de pouce" (Helping Hand) is an association under the law of 1901, a partner in academic success. Most of its activity helps children struggling with schoolwork and problems outside of school.

65 MOUVES, Movement of Social Entrepreneurs. Let us note that the term "entrepreneurs" cannot help but recall Schumpeter and thus the world of the heroic entrepreneur.

66 *Jugaad* is the Hindi word given to innovations that do not involve any additional expenditure of energy. This type of innovation originates in developing countries.

is an addition, an additive. This is based on an environment of growing abundance.

However, over the last few years, a form of innovation has developed, which defines itself by the removal of an object or a social behavior. One might give examples of car-less towns, pesticide-free agriculture, supermarkets without plastic bags, food without artificial coloring and short trade routes. These innovations lead to a de-institutionalization of the commercial apparatus for a new recomposition.

This new form of innovation, called innovation of withdrawal, implies ruptures, dissociations of existing links, and the loss of habits. We understand, for example, the progressive disappearance of textbooks from schools.

Two authors[67] have studied this mode of innovation, in particular in the agricultural world, on *no-till cultivation* which asks farmers not to plow certain areas of their plots to avoid the depletion of soils and to take environmental issues into account. Certainly, the principle is still based on the search for profitability, but by withdrawal of the item.

This withdrawal will subsequently structure human agricultural activities and social relations. In other words, if the theory of translation in the appropriation of innovation is still relevant, social interactions will focus more on decreasing intervention in certain aspects. It thus becomes essential to develop persuasive arguments for abandoning certain practices. It is not a matter of creative destruction in the manner of Schumpeter, because here it is the withdrawal which is the innovation and not its previous condition of existence.

Of more recent appearance is something called frugal innovation. Some see this as an epiphenomenon, arguing that it does not fundamentally challenge the nature of innovation.

Now, as with innovation of withdrawal, it seems that frugal innovation might, on the contrary, reveal something about our social transformations and will doubtless have an indirect effect on certain pedagogical options at school.

67 GOULET F., VINCK D., "Innovation of withdrawal", *Revue française de sociologie*, vol. 53, no. 2, pp. 195–224, 2012.

First of all, what is frugal innovation[68]? The authors define:

> "frugal innovation as the ability to do more with less, that is, to create at the same time more social and commercial value while economizing on precious resources such as energy, capital and time" (p. 31).

Frugal innovation came from a developing country: India, where the current lack of resources invites its inhabitants to innovate with the materials available. The example given by the first authors on frugal innovation illustrates this option: an Indian man goes to work every day by bicycle. The roads being very bad and covered with potholes and big gaps, his bicycle bears the brunt of these holes and he must pedal harder by jolting. He then has the idea to exploit the energy of these jolts by the installation of a dynamo which captures this energy and allows him to have a small motor, facilitating his travel with less effort. This innovation is called "frugal" as it consumes less materials and produces energy. However, such an innovation is a double-edged sword, as it may lead politicians to tighten the purse strings and resources, in the hope of exploiting the imagination and creativity of actors to the maximum if the latter wish to live in better comfort.

This frugal innovation, in resonance with current trends of economic crisis and the search for energy economy, rests on six principles:

– engage and iterate, that is, instead of waiting for inventions to come from research laboratories, here we begin with the clients, the recipients of innovation, carefully observing their behavior, even associating with them;

– boosting agility which responds to the trend of customizing objects, which the authors call "mass customization";

– imagining plural solutions through continual recycling of the materials used in producing new objects;

– shaping customer behavior, for example, by giving a bonus to drivers who use new, non-polluting cars;

– co-creating value with "consum-actors" who conceive, develop and share the products and services that they want;

– collaborating with innovative partners, that is, creating networks in the chain of production of the new object.

68 RADJOUR N., PRABHU J., *L'innovation frugale*, *Diateino*, Paris, 2015.

Without going too deeply into each of the principles, we can already raise certain questions as to the possible transfer of such innovation into schools.

The first principle runs counter to some assertions which stress that research is essential if we wish to develop effective pedagogy and teaching; the results of research will support certain practices.

The idea of consensus conferences, hosted by teacher-training organizations and universities, seems to sketch out a concept whereby researchers and actors control the production of innovation, and not just the researchers in their laboratory. In other words, applied research would no longer be relevant.

Researchers inform practitioners verbally of some of the results of their research work and hold discussions with education staff to find out what is understood, what might be integrated into their daily practice, on the condition of rethinking everything by the standards of professional practice.

To our knowledge, there has never been an assessment and a follow-up to these conferences. Now, frugal innovation questions the role of research and invites researchers to work right from the beginning with practitioners (the consumers). The definition of the research function thus changes. We will see this again in the chapter concerning the link between innovation and research-development.

The second principle is close to the idea of personalization of education, that is, not to abandon mass interventions while still ensuring that everyone should have the feeling of being taken care of with their particularities and their own representations. School currently poses the question of the individualization of teaching, which is not the same as individual instruction. The difficulty rests in the balance between considering every student in their individuality, as well as the students as a whole.

The third principle, already mentioned in the discussion on innovations in schools, rests on the idea that, to innovate, there is no need to spend more, either money, time or personal investment. This requires imagination from teachers who, for example, using available resources, find new uses for materials which are already present in the schools. It is a way of saying to teachers that everyone can innovate without needing to wait for more resources. We will see that it is not as simple as that, and that any new activity requires from the person carrying it out a greater investment in time, or even in stress.

The fourth principle mentions the idea that innovation always takes place in a two-way relationship where the innovator influences the behavior of the recipient and vice-versa. The new idea is that this is done in a very conscious manner, in the goal of changing and influencing the behavior of the recipients of the innovation. This is not a case of adaptation of the innovation to consumer demand, but of co-production where the innovator is transformed as much as the consumer. For example, an innovative teacher who up until now has asked her students to study individually puts in place a series of innovations centered on group work and frequent exchanges between students, both in and out of class. This innovation will unquestionably change the role of the teacher and the behavior of the students to develop competences in collective work. The tools of evaluating student works will include criteria on these new competences.

The fifth principle, that of forms of association of the recipients of innovation, is very interesting for the teacher, in that it takes into serious consideration the role of students or parents faced with innovation and with the innovator. Now it is rare to encounter an innovation that originates with the students and is developed by the students, even if it is taken up by the teacher. This question of the role of recipients in the initiative, in its possible evolution, has never been addressed regarding innovation in schools. However, students and parents doubtless have new proposals to make which might be translated into innovations within the school.

The sixth principle relates to actions already carried out as well, at the level of the Department of Education with its innovation databases, innovations available on the national server and at the academy level. On the contrary, it is rarer to see associations being systematically formed between innovators working on similar themes or in geographic proximity, in the form of networks which are solidly established and maintained. To our knowledge, this is done, but not in a systematic way and, as usual, with no evaluation!

We have tried to make a parallel between this frugal innovation and what it could contribute at the level of innovation in schools. See Table 1.2.

| **Entrepreneurial frugal innovation** | **Frugal innovation in schools and its specificities** |
|---|---|
| How to do better with less? | The question of what is "better" remains open in an acute manner in education. To define it, must we rely on international tests (PISA, TIMMS, etc.), which are incomplete and homogenizing?<br>"Less" is envisaged in a material sense: what will not cost more in energy. But if this latter is understood as the teacher's energy, it remains to be shown as teaching is a very demanding profession. |
| It does not mean a "low cost" approach but a local one. | Schools which are locally managed (primary and secondary) and exploit their proximity to local resources. |
| A change in paradigm. The 4 Ds: deregulation, decentralization, digitization and deceleration. | Local management: local recruitment of actors, local determination of programs (linked to the common base), reduction in central directives with adaptable programs and durations of teaching, to adapt to the profile of students and their parents. |
| Mass customization, increasing scarcity of resources and growing demographic diversity. | Construction of its own content and tools/pedagogic supports based on local resources. Exploitation of the particular contributions of students and their families. |
| Shaping the behavior of the client | Parents and children by highlighting the behavioral implications raised by the innovation. |
| Co-creating value with consum-actors, shaping client behavior (encouragement, influencing without giving the impression of manipulation), co-creating value (proposing ideas for innovative products with a platform of co-creation).<br>Innovation invents a problem for which it becomes the solution. | Collecting the ideas of teachers (deprofessionalization and reprofessionalization).<br>Collecting the ideas of students and parents.<br>Analyzing their feasibility for implementation. This demands a continuous dialog between students, parents and education staff.<br>Innovation reshapes the ways of thinking about school and anticipates possible negative effects in the future. |
| 5 indicators: costs, resources, reliability, responsiveness and agility (adaptability). | These 5 indicators correspond to competences defining the profile of recruitment for school staff.<br>Responsiveness is valued more highly than reflection which comes second in terms of time as |

| | |
|---|---|
| | well as importance. In other words, the reflexive practitioner much touted over the last 20 years moves into the background in relation to the one who acts and reacts first, which does not prevent subsequent reflection on this action. Action is valued over words, just as in the learning process which is currently recommended to students. |
| Collaboration with innovative partners. | Opening the school to its environment, intensification of social networks via the Internet. |
| Developing a culture of frugal innovation. | Initial and continuous training of teachers centers on their capacities for innovation in a universe of limited resources, particularly on the fundamental development of creativity. |
| Growth involving rupture (minimalist design), a new spirit of authority. | The inverted classroom is an example of this, where students receive video "capsules" to prepare at home for exchanges in the classroom. The role of teachers and students produces a new spirit of authority. |
| Recycling all along the value chain (green products reused and recycled several times). Ecodesign and circular economy. | No more textbooks updated for each new program, but "video capsules" updated and enriched according to a spirit of gamification. |
| Growing large-scale personalization of profitable products. | Taking ownership of the tools available: sharing between teachers and students (using equipment belonging to the local council and associations) |
| Sharing rather than buying. | Bricolage, ingenuity and trading forums for items such as books or pedagogical and digital material. |
| Faster, better and cheaper. | Shorter circuits of trade, recruitment and mutual assistance allow more speedy acquisition and carrying out of activities. |
| Not creating new needs, but responding to a well-defined existing need (real problems of the clients), starting from client needs: "the customer is always king". | It is hard for a school to create new needs; it responds to new needs corresponding to social demands and to behavior in contemporary society (creativity, the ludic aspect of learning, initiative, etc.).<br>This recalls Schumpeter who created new needs, but without the co-participation of customers in this new need, which is far from being the case in schools. |

| | |
|---|---|
| R&D: prioritizing quantity of innovations over quality. | Collecting all the ideas of actors and working on them to develop new practices: a marketplace of ideas in the school establishments would be interesting, on the condition that they would be exploited in common and followed by new activities relating to these proposals.<br>It is better to try an innovation and then abandon it than to do nothing for fear of failing. Multiplying the number of innovations allows us to banalize these approaches and leads to the development of professional innovative competences. |
| Using crowdsourcing and social networks. | Solicit loans from generous donors and financial partnerships which are concrete in time and equipment. |
| Being steered by the market (Just In Time). | Establishing periodic evaluations over a short cycle, which are immediately exploitable. |
| Governance group for innovations (portfolio of products). | Have a sorting committee for innovations/ideas at different levels of the educational system. |
| There are no revolutionary technologies, and there are only cutting-edge commercial applications. | The inverted classroom is an example of the use of ICT and must be considered based on certain conditions of use. Commercial operations are absent in schools, but an extension of the inverted classroom may, for example, produce a cutting-edge school. |
| Take inspiration from start-ups: make it simple, work quickly, find solutions, do not fear uncertainty. | At school, the creation of mobile open work teams on multiple innovations, ready to abandon or modify them. |
| Giving power to employees, using the power of the group. | Power is shared with students and parents. The role of teachers and administrative and maintenance personnel is reshaped according to a clear definition of competences and attributes for each function. Each person is identified in a group or category of staff. |

**Table 1.2.** *Conditions for establishing frugal innovation in a school*

In other words, we have tried to answer the following two questions:

– What would frugal innovation mean in education?

– What would a frugally innovative teacher look like?

If we return to the definition based on the authors of this frugal innovation born in India, we might say that it would mean the capacity to innovate rapidly, at the lowest cost and with very diminished resources, according to a sense of resourcefulness and "bricolage".

Some examples from education might illustrate, at least in part, what we might consider to be frugal innovation:

– free courses in the form of YouTube mini-videos (storytelling and games), which is now common practice in some schools;

– MOOCs (Massive Online Open Courses), that is, platforms of resources for educational purposes available to the public, based on different conditions of assessment and monitoring;

– the reduction of bureaucracy and levels of hierarchy through more transparent decentralization (see Chapter 3, section 3.5, on the educational policies for development of innovation in schools);

– giving power to employees (*Maker Faire*)[69], that is to teachers, to students and to parents, according to a well-recognized codification and explicit sharing of responsibilities;

– intelligent recruitment of teachers, emphasizing their competences in creativity and openness and their previous experience outside school;

– gamification, that is introducing activities for students which are essentially fun and motivating;

– platforms such as CoLearnr, free of charge, for collaborative learning online;

– sharing discussions (FutureLearn) between students (according to their level) and teachers whose vision for teaching work is to include a process of life-long professional learning;

69 Originating in the USA in 2006, *Maker Faire* is at the same time a science fair, a public festival and a reference event for innovation. It arose from DIY (do it yourself) culture and brings together innovators wishing to improve their work. These events take place in several capitals. A *Maker Faire* was held in Paris in 2017, at the City of Science.

– creating a (soft!) ranking according to the level of participation in exchanges (reputation, appreciation by peers of the quality of comments);

– micro-schools funded by crowdfunding, which does not yet result in subservience to a media outlet, a business or a multinational. We might also think of crowdsourcing for teachers and students, that is, the externalization of certain activities such as drawing on networks of knowledge, of creating sites which are not restricted to the school world, etc. However, this demands meticulous work on the ethical modes in effect.

# 2

# Innovation in Schools and its Social Forms

## 2.1. Innovation in schools as social innovation

In France, social and institutional preoccupation with innovation in schools began around the 1960s. This does not mean that there were no innovations in schools before then, but these new actions were not designated as such.

The word "innovation" began to appear in official texts at that time. After that, its usage spread along with educational reforms. But what does it entail? What is the nature of the innovation that is mentioned?

Although originating in social innovation, innovation in schools, due to the mission with which it is entrusted[1], is a different thing again. It has its own identity, which may be defined by a few characteristics.

Therefore, in what way is innovation in schools, although belonging to the category of social innovation, different from it? It seems to possess its own identity which might be defined according to four dimensions:

1 Even if certain people have tried to put it under the heading of profitability, as in a firm. LAVAL C., *L'école n'est pas une entreprise. Le néolibéralisme à l'assaut de l'enseignement public*, La Découverte, Paris, 2003.

– Its goals: it contains within itself a model of a future society, that is, aspirations which go well beyond what social innovation (responding to current institutional needs) is able to bear[2]. The issue is that of a desirable society, including strong social values. In distinction from social innovation, its goals are embodied in the institution created by society, and its activities respond to it.

– Its forms: it involves the explicit and central form of learning. It offers situations to facilitate the latter. Its components are at the service of learning in a manner closely linked to current conceptions of learning, conceptions which are continually questioned by research. The various modalities of pedagogy are, in themselves, timeless (in the sense that all the activities implemented center around the pedagogic relationship, which is in itself timeless). Vincent (1994)[3] officially defined this form as follows: "systematic organization of education as a necessary step, in a specific place and time, which establishes a sustainable mode of socialization in a given historical configuration". This definition, still operational, is, however, questionable in that current societies, particularly Western ones, have experienced a partial transformation from industrial society to digital society, where places and temporalities are redefined. Innovation no doubt plays an important role in this redefinition of the school's identity.

– Its functions: it is part of the regulatory function of school as social glue, while opening up social possibilities which reconfigure the profiles of its beneficiaries (for example, in schools, developing the competence of creativity among students, when previous generations had to avoid behaviors judged to be marginal, so as to faithfully follow the norms established by authorities).

– Its actors: it is the multiplicity of actors within the school according to established complementarity.

2 We believe that education is the basis of all decisions about the future.

3 VINCENT G. (ed.), *L'éducation prisonnière de la forme scolaire? Scolarisation et socialisation dans les sociétés industrielles*, PUF, Paris, 1994.

We may sum up this specification as shown in Table 2.1.

| Domain | Explanation |
|---|---|
| **Goals** | School is a special institution which, if it is not completely independent, is the responsibility of the State public service. Society assigns goals to it, whose importance goes far beyond a single generation. No citizen may be excluded from these goals. |
| **Forms** | School is an institution defined by its scholastic form. It has evolved over time, but has always involved adults facing a generation which must develop those competences that are defined by law. The scholastic form, as defined by Vincent in 1994, is doubtless partly applicable, and innovations play a role in this redefinition. |
| **Functions** | Innovation in schools focuses on two essential functions: socialization and schooling.<br>If such functions also apply to institutions other than schools, these are also educational institutions and are announced as such. |
| **Actors** | There are multiple actors involved, but their roles are predefined by the assigned central objective: developing the competences of the young people of a country. They guide innovation in their common desire for this to happen. |

**Table 2.1.** *Specificity of innovation in schools*

## 2.2. Educational transition as innovation in schools

The concept of educational transition appeared about 20 years ago in the wake of other transitions, such as ecological transition which includes energy transition, digital transition or demographic transition.

What might this term mean? Why does it now appear with such a positive and futuristic connotation, of the same kind which innovation in schools held some time ago? What distinguishes it from innovation?

First, educational transition seems more radical than innovation to the extent that it targets a profound structural modification in modes of teaching and concepts of learning for young people. This is a disruption, which involves a paradigm change in the way of envisaging school, whose founding principles rest on the values held by a desirable society.

These values are relatively common to all those who call for educational transition and who want to respond to a new society.

Thus, 2015 saw the creation of two associative structures under the banner of "educational transition":

– The first, called Lab School Network, emerged in 2015. This English-speaking network is open to all countries, with the idea of a profusion of ideas and proposals for a new kind of school, different from what actually exists. This association cites the work of Bernard Delvaux[4], which can be summed up in the following sentence: "School is more and more out of step with social evolution. It changes little and slowly, while around it everything is changing in an accelerating manner". In other words, the role of school is to keep pace with social evolution or at least to respond to it as adequately as possible. In his work, this author offers several scenarios for a school of the future.

One of the scenarios depicts an educational system composed of multiple schools, each with a different project, where competition is the keyword (we have already seen this in the different organizations each trying to make their school into a laboratory, as Lab Net School did in opening a different kind of school at the beginning of the 2017 term).

Actually existing schools would develop a therapeutic relentlessness, while strengthening the traditional form of schooling. For him, this is a dead end.

The school he would prefer would "maintain a common educational institution, while giving it goals and a form which differ substantially from present schools" (p. 6). His approach locates him within educational transition, as promoted by this association, in that it advocates two basic principles: emancipation and representation.

Commitments regarding school are not without ideological characteristics, and the vision is often idealistic. How, concretely, would this proceed? Is it in these "laboratory" schools that the initiators believe they will find the concretization of their ideas? The question of transformation of actually existing schools into schools of educational transition is an open one and will not necessarily come about through these laboratory schools, but

4 DELVAUX B., *Une tout autre école*, Pensées libres, GIRSEL-UCL, Louvain-la-Neuve, 2015.

through a realistic vision of progressive mutation of existing schools through teacher training, evaluation in the schools, dispersal of age-based classes, modifying coaching activities for teachers, competences of school leaders, etc.

Everything up to this point comes under the heading of desire and militancy, with the aim of bringing together the greatest number of people from different universes. Except, the heterogeneous gathering of people of good will does not make for a school whose educational objectives would be explicit, around activities for developing students and a common base of building coexistence.

This Lab Net School hosts many functions bringing people together as well as expert seminars throughout the country whose contents, admittedly diverse, are inclusive. Their definition of educational transition is nothing if not surprising to the extent that it cites such diverse classical pedagogues as Freinet, Montessori and Dewey, in a syncretism as lofty as their convictions.

At a presentation on 11 January 2017, one of the founders of the movement, Pascale Haag, presented a slideshow including, at the same time, the origin of the word "transition" and supporting statements from some pedagogues. It explained:

– the expression "educational transition" came into use at the beginning of the 21st Century, in the context of political changes linked to democratization in the former Soviet republics (Satchkova, 2003) or in Africa (Elansari, 2001);

– a semantic shift came about more recently: educational transition began to appear more often as an extension of other transitions (towards sustainable development or digital technology);

– many "different" schools (Montessori, Freinet, Dewey, Steiner, Sudbury, etc.), mostly private schools (but also within the National Education system), fall within the framework of educational renewal; 67 new schools at the start of the 2015 school year out of the total of 772 in France, and 93 at the start of 2016.

We are witnessing Belvaux's scenario 1, that of the fragmentation of the educational system.

After one of these meetings, the Lab Net School asked its members what words they would prefer to use to define a "school of the future". These are the main ones in decreasing order of importance: innovative pedagogy, inclusion, personalization, digital, partnerships, active pedagogy, evolution, future and lifelong education. This set of words clearly displays a strong desire for change in teaching practices different from the representation that these people have of them and say they experience or have experienced. The construction of this common representation rests on semantically unstable terms, whose meaning will doubtless need to be deepened in order to become useful.

The second structure is based on the Potentials and Talents Foundation[5], created in 2014, whose roots can be found in questions concerning children of high intellectual potential, as indicative of the inadequacy of current pedagogical practices for the diverse profiles of young people. The central idea is based on the personalization of pedagogy, that is, a pedagogy customized for each young person, taking into consideration their identity and their individuality. The site explains that it is "for all these reasons that the Foundation has encouraged a network of innovative schools which proclaim the personalization of education as their highest value, to join into a federation: FETE, Federation of Schools of Educational Transition".

This foundation is administratively "hosted" by the Foundation for School, recognized as a public-interest organization by a decree appearing in the Official Journal of 20 March 2008. Its mission is to "facilitate educational initiatives, to support them, finance them and oversee the creation and development of independent schools"[6].

This "hosting" foundation has developed a Charter for the foundation's schools, with indications for structures and pedagogy, of which Box 2.1 gives some extracts.

5 See www.potentielsettalents.org.
6 Quoted from the Website of the Foundation for School, www.fondationpourlecole.org.

**Structures**

Free schools, non-profit and duly declared to be so; this may mean kindergartens and elementary schools as well as secondary establishments whether generalist, specialist, technical or professional. These establishments will have to be set up in accordance with law and the Education Code in particular. Under current law, they cannot receive State aid.

Granting an educational Charter clearly defining the vision and the educational and pedagogic methods which characterize the school, allowing parents, teachers, students and funders to get involved in full knowledge of the nature of this educational project; putting this charter into action.

Ensuring follow-up of every child as a unique person and consequently determining the staffing levels required.

Recognizing the role of the parents of students, the first educators of their children; informing and consulting with them so as to ensure educational coherence and harmony between home and school.

For secondary establishments, providing evidence of real reflection on the question of diversity.

Welcoming children from all walks of life insomuch as both they and their parents undertake to respect the establishment's Charter in all points.

Having sources of finance which are identified, transparent and diversified, communicated to the Foundation systematically and on request for all sources exceeding 10% of the total.

Recognizing the right of the Foundation to carry out site visits to verify respect for this Charter and to provide it with records of inspections carried out by National Education.

**Pedagogy**

Providing students with the know-how and knowledge necessary to exercise their future freedoms and responsibility, expecting rigor and consistency in the use of direct and progressive learning methods, ranging from the simple to the complex, ensuring good assimilation of knowledge by students.

Forming the intelligence, the will and the sensitivity of the child in transmitting to them a sense of the true, the good and the beautiful; in doing so, we develop their critical sense, their self-knowledge and their self-confidence, their respect for others and for nature, a work ethic, the capacity and the taste to assume responsibilities and respect for their commitments.

Recognizing in teaching and in the whole of school life the equal dignity of men and women.

Teaching rules of social life; exercising authority in an atmosphere of shared confidence, mutual respect and the joy of learning.

Preparing children to enter the national community, particularly by transmitting to them the knowledge and love of French culture and prohibiting the promotion of any political ideology.

Preparing children to enter active life; helping them pass diplomas recognized at national or international level; helping them decide on their choice of career.

Teaching in French or, exceptionally, in a French regional language or a language of one of the Member States of the European Union as a working language in the schools.

**Box 2.1.** *Conditions of membership in the Foundation for School*

This Charter for schools includes nothing worrying, except perhaps its silence on any aspect of the quality and the level of professional training of teachers, of their career, of the role of continuing professional development and of the composition of pedagogical teams. The pedagogical section is more based on general principles than on practical options such as privileging group work. Such a Charter leaves the "hosted" establishment a lot of leeway in enacting these principles.

Using a new term does not always produce a new reality; facts are stubborn things. Novelty is not always new!

However, it is interesting to observe a growing movement to respond to the current social challenges which National Education is scarcely coping with; and the word innovation may be hard to maintain in the face of impatience and skepticism, in proportion to what has been tried for many years in terms of innovation, especially in that it remains sustainable and would be generalized.

## 2.3. How do we identify innovation in schools?

If we return to the National Education system in France and to public education, which, up to this point, means the majority of schools[7], we note that there is no accounting for the number of innovations actually put in place, while support for innovation has been installed in the academic infrastructure at all levels of schooling.

Why such ignorance? Why such a vacuum?

One cause is the difficulty of identifying innovations. And the definition of innovation in schools, which we have previously given in this work, will not allow easy identification of innovations.

In fact, it sometimes happens, during observations of classes, that we characterize a pedagogy as innovative while the teacher is not aware of innovating and would never have declared themselves to be an innovator. There are thus innovators who do not see themselves as innovators, because they have a more radical idea of innovation, while others come to tell them that they are innovating through practices, which might be considered minor or routine, already carried out in other institutions.

In one study, we asked two inspectors in the same discipline to design innovative practices. Once these practices were collected, we asked each one to consider the innovation of their colleague and say whether they would have placed it in the category of innovations. The rate of disagreement was 60%! In other words, what one considers an innovation, the other does not see as such.

This is because innovation in schools benefits (or suffers!) from a strong ideological charge in a field that is already ideologically charged (the goals of schools). Pelletier[8] refers in this regard to three types of psychic investment in innovation:

1) an innovation-religion with strong symbolic charge, which communicates great social desirability, a source of mobilization for people;

---

7 One high school student out of five attends private school. See http://www.education.gouv.fr/cid57105/geographie-de-l-ecole-douzieme-edition-2017.html.

8 PELLETIER G., Piloter une innovation ou l'art de gérer l'inutile, Conférence présentée aux deuxièmes journées du savoir vert, Ministry of Agriculture and Fisheries, Paris, 1994.

2) an innovation-project which retains a symbolic charge, but adds a managerial aspect of carrying out change which is planned, structured, directed and evaluated in a kind of "engineering of innovation";

3) "why-not" innovation which is a momentum of possibilities, freedom and the creation of life.

Similarly, Prost[9] considers that all innovation is carried forward by a kind of political or union activism, which means several things:

– "innovation is formulated in terms of democratic values: it is a political as much as a pedagogical project and is rooted in conviction;

– innovation is experienced through personal engagement, in permanent mobilization;

– innovation is carried out by groups which fulfill several functions. A technical function of reassurance and mutual enrichment: innovators compare their experiences and improve them. An internal sociability function: they create welcoming groups whose conviviality may be seen as an antidote to the bureaucratic nature of the institution, with all the tensions, rivalries and occasional splits which are associated with affective investment. An external recognition and legitimization function: in forming a group, innovators create a collective existence, in the face of administrators and colleagues;

– innovation is seen by pedagogic activists as the solution to the defects of the educational system. Innovation is a response to concrete difficulties or to flaws in the system, which can no longer be tolerated;

– innovation creates a repertoire of alternative pedagogic practices. Successful innovation is that which becomes a pedagogical custom."

We now understand more about the issues of innovation in schools and the difficulty in understanding and evaluating it! Yet, the currently recurring question is that of the effectiveness of innovation, which is sometimes attributed to its means or its allowances.

9 PROST A., L'innovation, intervention at the Educational Biennale in Paris, 1996.

As soon as innovation enters a process of identification, it seems to lose its spontaneity and its uncertainty. To identify an innovator and their innovation, the latter must clearly signal its qualities and its profile so that it can be sold. The process of communicating an innovation is thus at work. How to identify an innovation without having been made aware of it and without having been introduced to it? As the saying goes, "to live happily, live in hiding". Some innovators do not want to expose their innovation to inspectors, to the administration and/or to media of whom they are wary.

This identification is easier when it concerns an innovation in structure, in organization or even in the space of partnership, to the extent that it is essential to obtain formal permission for it. For a pedagogical innovation, at the very heart of the teaching profession, such identification is less easy. If the innovation involves a pedagogic team, as preferred by the authorities (some of whom state that all innovation in education is collective), innovation carried out by a teacher in class will be scrutinized for better or for worse by the administration.

This current vision of innovation overlooks ingenious pedagogical experiments or bricolage by a single innovator. If the latter innovates and wants recognition, they must do so in a team and preferably in a multidisciplinary team. If it involves a partnership, this corresponds better with currently identified and celebrated innovations. This organizational process requested by authorities is based on the result of recent research that deals with the collaborative construction of knowledge and the enrichment of thought and knowledge through creating a "collective intelligence".

To this, we can add that, currently, new ideas about schooling and transformations rely on modern conceptions of Darwinism, stemming from a current inspired by biology and studies of development in health and its support. Basically, school is urged to respond to social and technological evolution with rapid adaptation.

Innovation, however, is deeply embedded in the aporia of its identification: who declares that a given educational or pedagogical action is an innovation? Currently, research in cognitive science leads us to shift away from a psychology of learning centered on the development of the young person, towards the idea of the existence of a "smart school", in the same way that there are "intelligent" towns or "learning" organizations.

Remember that since the 1960s, in France, innovation has been the object of bargaining at every institutional level: whether a head of school wishing to improve their school and pushing a team to innovate so as to have something to promote at the academic level, or an inspector wishing to show that he is making changes or a principal being proud of their policy of change at school.

Since then, a series of central and board of education-led mechanisms have been put in place to try to collect innovations, capitalize upon them and transfer or assess them. Innovation in schools is therefore from the outset involved in social exchanges between different actors who determine whether it is an innovation or not. Thus, we return to what authors like Akrich, Callon and Latour said about innovation in business.

In 2005, an institutional filter for collecting innovations in schools was put in place through a Department of the Ministry of Research, Innovation and Experimentation (DRIE). Two words had been attached to innovation, namely research and experimentation, in the hope of proceeding to support that is strengthened by researchers, not allowing innovations to appear as if it would be hard to define boundaries or to think of possible transfer.

This filter indicated that every innovative team would need consent for its activity by the Board of Directors and the Teaching Council of the school. As soon as such agreement was obtained, a whole series of procedures and documents needed to be filled to allow the team to obtain allowances of hours of leave, use of school time, new partnerships and sometimes financial means. The innovative team would thus be listed by the administration and might benefit from support, whether from researchers or trainers.

More recently, inspired by already-established institutes where a laboratory or group of laboratories cooperate on a prototype aiming for transfer of technology, we saw the development of the Camot network, a federation of businesses, research units and multidisciplinary experiments (see Chapter 3). Can we imagine poles of competitiveness in the educational field?

In addition, a reflection group was formed around a researcher in medical studies, mandated by the Ministry of Education. This group published its report[10] in accordance with the terms of reference given to it by the ministry. At the same time, another report, also official, from CNIRE (the National Council of Innovation for Educational Success)[11] appeared with a series of recommendations. These two governmental initiatives were carried out in parallel with the activities of CARDIE (advisers in research and development, innovation and experimentation).

## 2.4. Innovation in schools and "best practice"

"Best practice" has often been assimilated to innovations considered successful and likely to be effective. But what does this mean?

"Best practice" emerged in the international context. The term *best practice* implies applying a gradient of value to the category of "better", not to mention innovation as also a supposed bearer of improvement.

This emergence is in strict relation with the concept of "benchmarking" – competition and comparison between national educational systems, so as to identify the best of what is being done in education. Behind this concept lies a concept of development of educational systems through a technocratic aspect and through management. In fact, "best practice" carries an undertone of a form of transfer, where it would be enough to demonstrate activities to teachers for them to develop "best practice". Now, we know that professional practice, and even more that which is involved in the pedagogic relationship, strictly depends on a multiplicity of situational factors and parameters.

Note that when we speak of "best practice", we must still define the means by which it is communicated! How do you communicate a practice? Most often, this practice is described by an outside observer, or even better, by the actor themselves. This description may be written in a narrative form, on video or orally. It is not the practice that is communicated, but the formalization of the practice. This formalization greatly depends on who is

10 TADDEI F. *et al.*, Vers une société apprenante. Les grands axes, Report submitted to the Ministry of National Education and Higher Education, Paris, March 2017.

11 WATRELOT P. *et al.*, Innover pour une école plus juste et plus efficace, Summary of the work of the National Council for Innovation and Educational Success (CNIRE), Report submitted to the Ministry of National Education and Higher Education, Paris, March 2017.

doing the formalizing and on their intentions, which may be promotion, embellishment or popularization.

In other words, a practice does not spread to the extent where it is appropriate for every situation, in all its complexity and its intersubjectivity (it is rare for there to be a student who testifies to a teacher's practice).

"Best practice" is considered as an example to follow, a model, a reference, an "ideal type" and a successful experiment. It consists of innovative practices, which have been tested and evaluated as successful.

Guijarro and Raimondi[12] consider that "'best practice' is an approach which is often 'innovative' and which has been tested and evaluated and whose success can be presumed".

The two concepts are linked. As principal characteristics of innovations, Guijarro and Raimondi[13] suggest considering them in their local context, while "best practice" is better considered with the intention of generalization.

As with the identification of innovation previously discussed, we may legitimately pose the question: who designates this "best practice"? According to what criteria? Because the moral qualification of "best" suggests the existence of worse practice and therefore of censors anxious to make a judgment of value.

Through a semantic shift, this "best practice" is often reified, becoming those procedures and techniques that correspond to references and criteria in a syllabus of good activities. This standardized conception of pedagogic practice is still questionable, because, even if this "best practice" is evidence-based, this is not a guarantee that a practice will perform well in whatever context. If in medicine we can suppose that the relevant techniques can be administered, in education we are far from such an application. Also, research in education is plural and uncertain, which hinders an evidence-based policy.

---

12 *Développement curriculaire et bonnes pratiques en éducation*, UNESCO, BIE, 2003.

13 GUIJARRO B.R., RAIMONDI G., *Estado del Arte Sobre las Innovaciones educativas en America Latina*, Convenio Andrès Bello, Santiago, Chile, 2000.

What are the objectives of this collection of “best practice”? Here are some of them:

– make known new ways of doing things;

– make teachers want to change their practices;

– train teachers;

– create a database of successful professional practices;

– construct a professional syllabus based on the practices and competences that they are supposed to pursue;

– improve the performance of teachers;

– establish criteria for evaluation of teacher practices.

In 2016, a report from the Inspectorate-General (IG) of National Education included “best practice” in its title. The authors recalled the managerial origin of the term and insisted on the particularity of pedagogic practice: “in pedagogic matters, confusing procedure and teaching may result in little learning and a lot of damage. The effects of teaching also reveal themselves over a long period which further complicates their appreciation” (p. 4)[14].

This report clarifies the scope of a possible survey of “best practice”:

> “The Mission agrees with the conclusions of Sweden’s National Education Agency[15], for whom best practice cannot claim to represent a solution, but is aimed at proposing a range of possibilities which may inspire action adapted to a precise given situation. Thus, the conditions of realization which define practice forbid the construction of pedagogic models and any temptation to transform them into recipes.”

14 Le recensement des bonnes pratiques dans l’enseignement professionnel, Report no. 2016-078, IGEN, 2016.

15 Source: BRASLAVSKY C., ABDOULAYE A., PATIÑO M.L., Développement curriculaire et “bonne pratique” en éducation, UNESCO, BIE, Paris, 2003.

However, the IG has determined the criteria to select "best practice", which are mentioned in the report, namely:

– the coherence of the whole of what is in play in a moment of teaching (project, approach, progression, the place of the session in sequence, activities and goals);

– the behavior of the teacher, their manner of speaking, their written records;

– supporting students in their individuality, the multiplicity of micro-signs which are sometimes unintentionally addressed to them, but which they perceive precisely, and which may influence their activity.

Take what you will from this, as these criteria are of such complexity and subjectivity that, if the choice is not random, it could not be more uncertain.

In other words, the notion of "best practice" is a management tool for an educational system policy, in the sense that it consists of highlighting certain accounts of practices which are considered by those in institutional power to be exemplary and thus a source of information for those training or wishing to become high-performing teachers; a sort of calibration of efficiency as a path to follow! Here, we are very far from innovation in schools.

## 2.5. Promotion and transfer of innovation in schools

### 2.5.1. *Promotion*

We are still faced with the burning question of promotion of innovation, which links indirectly with that of "best practice". If institutional leaders have over the past half century tried to take innovation in schools into account, it has been with the intention of making it play a role in the evolution of schools.

But how to make this innovation known? How to promote it without rejecting any practice not characterized as innovation?

Several parameters allow us to determine the possible form of promotion of an innovation and its effects:

– the level of the innovation: that is, whether it concerns an innovation within an establishment, in a class, by a teacher;

– the actors involved: a teacher with their class, several teachers with a class or several classes or even external actors like a local authority or parents;

– the scope of the innovation: if it involves several classes in the same establishment or several establishments or other institutions;

– the goals of promotion: that is, whether it is simply a matter of motivating the necessary actors towards the proper conduct of the innovation or whether it is more about indirectly addressing the people concerned, or coordinating with the construction of initial or continuing staff training modules, or getting involved with a process of capitalization of innovation;

– the form of address of the promotion of the innovation: advertising, involvement in a tool for capitalizing on innovations at the local, academic or national level, the enlistment of higher hierarchical authorities such as pedagogic advisors or inspectors;

– the form of promotion: that is, the media used such as posters, written storytelling, participation in seminars and involvement in training activities.

We have tried to summarize all of these cases in Table 2.2.

| **Level of innovation** | **Scope** | **Actors involved** | **Goals of promotion** | **Form of address** | **Form of promotion** |
|---|---|---|---|---|---|
| A temporary pedagogic practice | In a class | The students of the class and the teacher | Inviting colleagues within the school to participate in or at least share the new goals assigned to this innovation | Orally, in meetings of the Teaching Council and informal meetings of teachers<br><br>Invitation to visit | Preparing arguments for an oral presentation of conviction and social influence |

| | | | | | |
|---|---|---|---|---|---|
| An activity on several levels | Concerning a level of the school | Teacher teams, the head of the establishment | Obtaining the support of the maximum number of teachers in the establishment | Interventions in the elaboration of the school's project, at meetings of the Teaching Council and parents<br>Production of a written fact sheet relating to the establishment and execution of the innovation | Oral argument adapted to the target audience<br>Creating an attractive internal fact sheet, a short video to be shown at meetings |
| An action involving a whole establishment | The organization as a whole | Teacher teams, general staff, administration personnel, the head of the establishment | Involvement in the establishment's project with the Board of Education's approval to benefit from financial assistance and possible allowances (CARDIE) | Production of a formatted document leading to registration at the regional or even national level, to benefit from support with training and research | |
| Activities in several institutions | The actors in the institutions involved | | Involvement in the projects of the institutions with the Board of Education's approval to benefit from financial assistance and possible allowances (CARDIE) | | |

| | | | | | |
|---|---|---|---|---|---|
| An innovation concerning all establishments at the same level (elementary schools, high schools, etc.) | Inspection or several inspections | | Involvement in the establishment's project with the Board of Education's approval to benefit from financial assistance and possible allowances (CARDIE) | | |
| An innovation at national level | National leaders and Academy principals | | Involvement in the establishment's project with the Board of Education's approval to benefit from financial assistance and possible allowances (CARDIE) | | |

**Table 2.2.** *Parameters for the promotion of innovation in schools*

### 2.5.2. *Transfer of innovation*

The transfer of innovation in schools is a complex process which supposes that we can detach an action from those who performed it and, above all, from the context in which it occurred.

Philippe Perrenoud[16] emphasizes that the idea of transfer is often easy for our leaders and for some teachers. We sum it up in Figure 2.1.

16 See www.unige.ch/fapse/SSE/teachers/perrenoud/php_main/textes.html, accessed 5 April 2017.

This schema rests on the idea of a movement in space, as the metaphor of transfer suggests, but complexifying it, in that passing from practices to representations of these practices, followed by their translation into new practices, is not done in an obvious manner.

The dialog (whose form of communication would be pertinent to define, as Table 2.2 does, namely as written, oral, pictorial, informal, video, etc.) to create an idea in the mind of B, which will inevitably be different from that of A, to then be modified in practice, is far from being simple and easy. Let us add that this dialog may be that of the deaf or be facilitated by a particular teacher or network (this might be an association, a group of staff in continuing education, etc.).

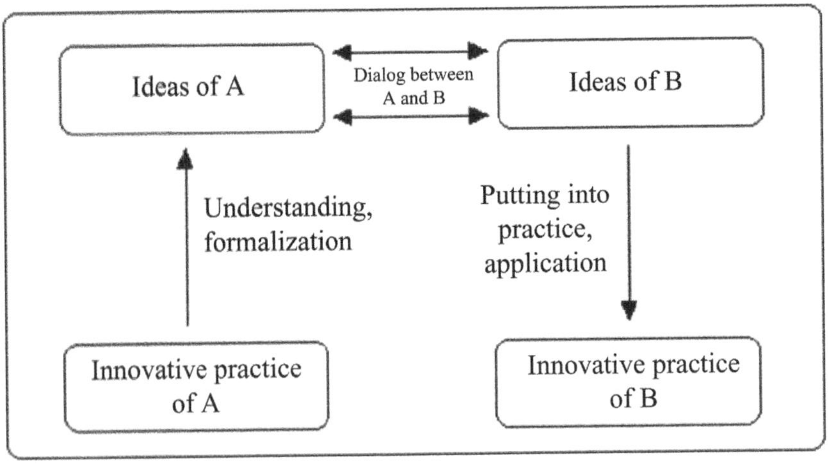


**Figure 2.1.** *Simplified schema of the transfer of innovation in schools*

The transfer of innovation in schools is incontestably part of the paradigm of communication. An undisclosed innovation only exists locally and is at risk of disappearing with its actors who, themselves, do not last forever. The problem is that many innovations are strictly dependent on their authors, according to a paternity inseparable from its designer.

The transfer of innovation is the cornerstone of the development of all innovation. Once it is identified according to various processes, innovation must be communicated and therefore communicable.

This communication obeys the rules associated with any communication, whose requirements we must remember. We have already mentioned this question with regard to the training of adults[17], particularly teachers. All training is based on intersubjective communication: it invites the exchange of representations, modifying views and opinions, engagement with other methods of professional action.

Most of the time, this communication occurs through verbal language as well as sign language or body language. This communication may use different media: this can be an informal conversation in a corridor during professional training, or an exchange of tweets or videos. The time of communication may be synchronous (interview, telephone, online chat, etc.) or asynchronous, as when it involves filming footage of a pedagogic innovation and putting it on YouTube[18].

Communicating seems to be a banal, ordinary activity. The practice of this activity often leads a person to assume that their knowledge about communication is relevant, knowledge which may take the form of a personal theory of communication, even to the point of a sometimes naïve vision: if an innovation does not develop, it is because it was not adequately communicated! Such a simple answer does away with the relationship of forces, power relations and the issues of communication. As Bourdieu[19] notes, "we must take care not to forget that the preferred relationships of communication, which are linguistic exchanges, are also power relations which actualize the balance of forces between the interlocutors or their respective groups".

17 CROS F., MEYER C. (eds), *La communication interpersonnelle en formation*, Chronique sociale, Lyon, 2014.

18 Here we can mention the role of mirror neurons, which, if they are at work during direct observation in class of a pedagogic innovation by one or more teachers, may produce future changes in the practice of the observer. In fact, mirror neurons are those brain cells which show the same activity when an individual performs an action as when they see another individual performing the same action, or even when they imagine such an action (fiction), thus the term "mirror". See: RIZZOLATTI D., CORRADO S., *Les neurones miroirs*, Odile Jacob, Paris, 2007.

19 BOURDIEU P., *Ce que parler veut dire. L'économie des échanges linguistiques*, Fayard, Paris, 1982.

We thus understand that when an inspector or a director from the ministry communicates to a group of teachers the benefits of an innovation, suspicion may arise among the teachers concerned, who may feel themselves judged as professionally incompetent or manipulated in their day-to-day practices, which are seen as invalid.

Here, we find one of the aporia of innovation: how does it happen that innovations which have proved themselves are not taken up by other teachers? What causes this blockage?

The channel of communication is very important, and in the last few decades, the Ministry of Education has done a lot to make innovations known and to capitalize on them on the official site Expérithèque, as the Ministry of Agricultural Education has also done on their site Pollen. However, no studies exist to this day on the use made of these sites and their concrete impact on the development of teaching practices, on the one hand, and student learning, on the other hand.

The mode of presentation of innovation is important, because even if it develops in the form of video films, it remains bound to a form of presentation. For some examples of communication of an innovation, we will give an extract from a written document concerning an innovation which has been highlighted and promoted by the Inspectorate (Boxes 2.2 and 2.3).

*An orchestrated set of practices at Maryse Bastié Professional High School (Hayange, Nancy-Metz Academy), France*

The high school includes three streams of training from CAP (Certificate of Professional Aptitude) to professional high school diploma:

– food and hospitality professions (around 160 students);

– business and customer service professions (around 120 students);

– support, care and personal service professions (200 students).

It also includes a 9th-grade preparatory class for professional training (24 students) and a disabled student's inclusion unit (10 students). The 2014–2018 school project, designed by the whole of the school community (educational teams, general staff, students), was developed in the context of reaction to two years of appreciable decline in exam results.

It includes four axes:

– axis 1: a benevolent and welcoming school, a site of flourishing;

– axis 2: a school which is a place of personalized training, supporting students towards successful professional insertion;

– axis 3: a school which shapes the responsible citizens of tomorrow;

– axis 4: a school which interacts with its cultural, economic and international environment.

This collective work, driven by the management of the institution, has been able to give form to different goals, deployed in a series of pedagogic initiatives at different levels, which may range from the class to the whole establishment. The mission has thus been able to discover:

– specific support for the 9th-grade preparatory class through team training, translating in particular into a professional half-day per week, which allows the design and enactment of projects, meetings with professionals, immersion in other professional high schools in such a way as to help students plan their future;

– use of student aid tools such as booklets of professional competences, generalized in the CAP classes, which welcome not only students with disabilities, but also all priority groups (students with learning difficulties or non-French-speaking students);

– personalized support, followed up through the levels, involving a key person in the management team, and consultation meetings allowing teachers to get involved in diagnostics and follow-up for every student;

– diffusion of a pedagogy of projects, implemented across the whole school: every student entering the first year of CAP, whatever their course, does a micro-project for the November–December period and another in March–April.

In each project, the students aiming for the professional diploma in Customer and User Reception and Relations are mobilized for organization and enactment. For the 2015–2016 school year, a project piloted by two secondary professional education teachers (CAP and professional diploma), based mainly on classes in the first year of CAP Food Service and the second year of professional diploma in hospitality, was titled "Fighting against food waste". It involved the whole school community, in strict collaboration with the school lunch service and the teaching restaurant:

– tutoring for the professional diploma in Commercialization, Food Service and Cuisine, and through professional tutoring, in the presence of teachers, final-year students train their colleagues in the class below them;

– a school newspaper (MAG) and a radio station broadcasting live internally three days per week, directly integrated into classroom practice;

– a European section involving opportunities for employment in Luxembourg, offered in the diploma of support, care and personal services, comprising 14 volunteers, with the obligation to speak Luxembourgish (4 hours of language per week, practical care work in a non-linguistic subject, periods of training in a professional environment in Luxembourg).

Thus, an overall strategy is worked out, taking account of the pedagogic context and goals, by each pedagogic team, integrating professional teaching and a section of general education. It facilitates learning and allows students to discover the meaning of the training. Every team sets out, at the beginning of the training cycle, a pedagogic project, which includes, over the two-year (CAP) or three-year (professional diploma) cycle:

– planning for training time in the school and in a professional environment as well as periods of internal assessment;

– disciplinary and interdisciplinary progressions, starting from analysis of the syllabus;

– the goals pursued in the different projects.

This collaborative set of practices greatly advances the team's work. It is based on the proactive impulse of the management team and calls upon the inspection team to support and guide the pedagogic approaches.

It mandates the meetings necessary for collective work. The guidelines on use of time are as follows, for example, for managing personalized support (PS) for the professional diploma:

– teams formed in June from volunteer teachers, which meet at the end of the school year to be operational at the beginning of the school year in September;

– two hours of PS per week, aligned in strips of time by level, all students of the same level being mixed together without regard to their stream;

– follow-up at every level under the guidance of the management team (the head teacher for the first year, the head of operations for the second year and the deputy head teacher for the final year).

Indicators for each of these activities are still lacking. For the head of school, as for the inspectors who accompanied the mission, the first criterion for evaluation concerns the climate of the school, and thus axis 1 of the establishment's project, clearly realized, as confirmed by interviews with students and relationships with the parents of students, frequently called to meetings and attending. The school is now recognized by its partners: links with professionals are facilitated, which cannot help influencing internships and the professional insertion of students. Impressive press coverage over the years 2014–2016 (18 long articles in the local press describing projects and days at school between September 2015 and June 2016) also shows that, in terms of the establishment's strategy, the goal has been achieved.

Other criteria of success, more directly linked to the immediate interests of the students, may however be identified:

– since 2014, exam results, which had been significantly down over five years, have improved;

– concerning support for the 9th-grade preparatory class, the school has achieved a 100% rate of students going on to 10th grade at their first attempt;

– the mission's interviews with students show that they are generally aware of the marked pedagogic difference right from their entry to the school: "we have more freedom"; "they take us more seriously, like future professionals", "they want us to develop; at junior high, we were robots, you work, you learn, otherwise we go on to something else".

**Box 2.2.** *Extract from a written account of innovation at a professional high school*

***Reconciling students with learning through professional situations, Louise Labé High School (Lyon, France)***

**The establishment**

The Louise Labé professional high school is labeled as a school for careers – textiles network, leather fashion. In 2015–2016, it welcomed 385 students with a high proportion of girls (67%) in connection with the school's offer of tertiary-style training. The latter is broken down as follows:

– A 9th-grade preparatory professional class.

– Training leading to the following diplomas:

- professional diploma in fashion and clothing careers;
- professional diploma in management and administration (M&A);
- professional diploma in commerce;

- professional diploma in sales;

- professional diploma in customer and user reception and relations (CURR).

**The context of the observed practices**

The M&A professional diploma was initiated at the beginning of the 2012 school year. Its syllabus offers an introduction to representative professional situations in the career of an administrative worker. The pair of teachers in charge of professional education for 10th-grade management and administration made the choice from the outset to prioritize, among the recommended pedagogic methods, an approach based on establishing scenarios of activities confronting students with several professional situations over a single sequence, thus reproducing real conditions of everyday professional activity. For their part, three teachers in charge of professional education for 10th-grade CURR intervened in 2014–2015 in a 10th-grade class, which was perceived as particularly difficult in terms of class management and disinvestment in learning. The students were experiencing particular difficulties relating the teaching delivered in school with the reality encountered during work experience. This situation gave rise to an ambition to change their practices, relying particularly on the experience of their colleagues in Management & Administration. To this end, they made the choice to create a professional context from scratch, and thus learning scenarios for their students. The latter would be running a camping ground. To this effect, they designed the following pedagogic Website: http://louiseenardeche.e-monsite.com, which includes all the resources allowing them to cover the entire syllabus.

**Goals**

By putting the observed practices into action, the teachers aimed in particular at the following goals:

– contributing to the ownership by students of training, which they have not always chosen themselves (M&A in particular), and in any case, to their knowledge of often unheralded careers;

– offering new sources of motivation by giving more meaning to learning;

– progressively building professional competences through activity in a real situation;

– developing professional behavior adapted to the contexts and situations encountered.

In the end, it involves the teachers allowing their students to regain confidence in themselves, to open themselves to new contexts and to establish new relationships with their peers and with relevant adults.

### Approach

The observed sessions took place in half-classes and for each of them, in a room reconverted into a professional space consisting of an open-plan office with workstations grouped into circles, a reception area and a meeting room equipped with a video projector. The scripting of activities is built upon models of close functioning in the two specialties. Entry into the classroom takes the form of rituals, in M&A in particular: prior seating of students in charge of reception, greeting the teacher with handshakes, signing in and sitting at workstations. Split into different services, students receive their roadmap (daily in M&A, weekly in CURR), which shows the activities that they must carry out while autonomously organizing. Resources, mostly digital, are put at their disposal to this effect. Depending on their positions and on the degree of progress of the mission, some students are in a phase of action (making a quote, contacting a client, etc.) and others are in a phase of reflection (analyzing the structure of a mail merge, establishing criteria for the choice of suppliers, etc.). The teacher is present to support their work and to proceed in the necessary corrections. They will sometimes have an explanatory interview with a student while their colleagues are in action, facing their workstation. The teacher's intervention thus contributes, through their questioning, to the acquisition of the competence being worked on. Thus, in M&A, during the session, the teacher, playing the role of the head of services, called a meeting intended specifically for a few students whose current tasks were linked to setting up payslips. This meeting was a chance for the teacher to, at the same time, lead the students in analyzing the work being carried out and to introduce new information, explaining the latest legal developments concerning payment for overtime. The teacher is the head of the orchestra who, for example, during the session observed in 10th-grade CURR, organizes a telephone call from a student attached to customer service to a client, then subsequently leads its co-evaluation by peers who observed it. In this model of functioning, students change jobs according to a plan (whose frequency may be bi-monthly, weekly or even shorter), which takes account, to some extent, of each one's acquisition of competences. Depending on the nature of the scenarios and the missions entrusted to them, the students are led to interact with their peers or even to train each other. In M&A, administrative simulation software was designed at national level to enable the creation for each student of individualized work situations. This simulator works with integrated management software and different work modules: electronic messaging, collaborative work space, supplier sites; time management tools; project management tools; electronic document management

> tools and online banking. The effects are observed. The observations highlight in each case the calmness introduced to the working environment. The students showed clear evidence of autonomy. Exchanges with them clearly show their good understanding of the activities they were carrying out. They were capable of explaining them very clearly and, furthermore, with pride. They visibly showed pleasure in working close to professional reality and in the ways suggested by their teachers. They particularly appreciated collaborative work and mutual assistance between peers. They were conscious of the progress they had made since the beginning of their training, and were confident in what they would accomplish from now until the end of the year. The teachers were equally satisfied with the implementation of these practices, which allowed them to attain the fixed objectives. They did not envisage "going back". The CURR teachers noted a significant decrease in absenteeism.

**Box 2.3.** *Extract from a written account of innovation in a trade school*

Innovations in schools are often communicated by fact sheets framed according to identical rubrics, which allow, as in the preceding examples, the innovation to be put into practice according to traditional criteria. These criteria are often difficult to square with innovation, which cannot be an action but is programmed and sequenced, or with evaluation, which is almost complete before its enactment.

We have already seen that these fact sheets respect the following framework: context, objectives, execution of the action and its evaluation. This routine may dissuade an innovator from filling out a sheet which is so standardized, but gives leaders, in turn, the possibility of evaluating this new action.

For about 10 years, the ministry's official Website has featured a database of innovations[20]. Its presentation is similar to the two examples given, but we find more and more videos that seem to better respond to the uniqueness of an innovation, which is hard to translate into writing. How do

20 This site is called Expérithèque (eduscol.education.fr/experitheque) and identifies more than 5,000 innovative and experimental projects implemented in schools and educational establishments. Its goals are to communicate with the general public and to leaders of national education; supplement ministerial communication with the general public and the media and make visible new modes of organization and new professional practices which are relevant with regard to ministerial goals to improve the performance of the public service.

you pass on an innovative practice through a written message? And under what conditions?

Let us add that this ministerial site insists on the choice of innovations as one of the tools to influence schools towards the desired policy. A few years ago, a Website was set up (hosted by the Board of Education of Toulouse), which had a webmaster, but whose content was provided by the innovators themselves. The latter would communicate their innovation as they understood it and provided their details to the consultants, with an invitation to come visit the site of innovation.

The webmaster's only job was to regulate and filter any innovation which would not fall under the teachers' code of ethics.

Why this constant need to centralize everything and to direct innovations, valuable or otherwise, by means of the strong arm of CARDIE towards the ministry? Innovators sometimes get the feeling of being under surveillance. Every innovation is in some sense a transgression; how then to reconcile acceptance of central authority with innovation? The answer doubtless lies in the development of autonomy at different levels of the educational system.

We must yet note that the profile of the ministerial site for promotion (or diffusion of information) on innovations/experiments has benefited from wider development over the last 5 years. Advances in digital technology have provided the possibility of putting texts online, which are richer and more narrative. For more convenience, in this process of dissemination, a standard framework is always available. Box 2.4 is an example of an innovation appearing on the ministry's Website in the "Expérithèque" file[21].

**Louis Timbal Junior High School**

5 avenue Jules Ferry, 87290 Chateauponsac, France [22]

Louis Timbal is "a learning school", welcoming and working with students from family and social profiles and backgrounds as different as they are numerous. As the public evolves, so does our work. In addition, self-esteem, well-being, a sense of belonging, good knowledge of oneself and one's personal and cognitive qualities are at the heart of a philosophy which affects both students and staff. This is an approach based on good sense and goodwill as well as good humor.

21 See http://eduscol.education.fr/experitheque/consultFicheIndex.php?idFiche=12845

22 Author: Fabrice Laurencier, http://www.clg-timbal-chateauponsac.ac-limoges.fr.

**Value-added action**

*New Study* is an initiative that allows us to really optimize study time, in particular in a rural zone with "captive" students, while making the work of AEDs[23] more valuable. It obviously requires a rethinking of classroom organization and student management. Important innovative work has been carried out, always with the goal of a sense of belonging and well-being at the school on the level of citizenship, its exemplary activity being the creation of a Council of Student Life (CSL). In four years, the CSL has become a well-known institution, recognized and now a pillar of the culture of citizenship in the establishment. It has been elected for three years by universal suffrage over one round with a list system. Each list includes eight names and respects the principles of parity and representativity (from 6th to 9th grade). Each list must also produce a poster for the elections which occur in October (School Democracy Day) in liaison with the DAVL[24]. For each student, the student diary serves as a voter card (a special page has been created for this), and two polling stations are available for voters outside class. These stations are managed by students and are part of the framework of school life. All staff of the college are also invited to vote.

In 2014 and 2015, three lists were presented; the participation of students ranged between 85 and 90% and that of staff around 50%. In 2016, there were four lists presented to the voters (32 out of 158 students, or 20% of the student population). Student participation reached 98% and that of staff reached more than 70%. The CSL includes 12 student members, four co-opted by the head of school on the recommendation of the school community from the losing lists. The students of the CSL propose a program for the year (presented to the Board of Administration) and carry out the following actions on it: the Day of Commitment, Christmas at Timbal, Carnival, "Timbal Days", the "Timbal Party", the Olympics, the Open Doors and a line of T-shirts and sweatshirts. They participate in writing and revising the Charter of Timbalians. The students of the CSL sit on the "menu" commission and represent the college to the outside world (commemorations at the local council, ceremonies at the Board of Education, etc.). They are equally the "showcase" of the establishment, creating every year a card which is the official greeting card of Louis Timbal. The activity of the CSL is a full participant in this research on the well-being of students at the college and, in turn, on their engagement in their capacity as students. The activity of the CSL is thus totally complementary to *Positive Re-entry* and to the activities involved in it.

23 The AED is an educational assistant, a contracted civil servant, recruited to public schools to provide assistance to the teaching team under the direction of the authorities responsible for organizing services.

24 Academic Delegate to High-School Life.

*Number of students and level(s) concerned*: 159 students from 6th to 9th grade.

### In the beginning

In our small rural out-of-town college, which has just emerged from a complete restructuring over more than two years, the student body is very diverse. Students encounter difficulties of method in organizing their work and in memorization. Students lack self-confidence and self-esteem. Required work is not always completed. They prefer to remain in the local area to pursue their studies, leading to little scholastic ambition.

### Goals pursued

We wanted to rewrite an overall project giving priority and dominance to well-being, self-esteem, the feeling of belonging and methods of learning, whereas previously, success in exams and orientation were our top goals. This is the philosophy of a health-promoting school, which we have progressively adopted. Students and staff, placed under these favorable conditions for well-being, should increase their effectiveness. Teachers should also seek to individualize methods of learning, so as to better target the functioning of each student. It seemed fundamental to us to let students and adults be happy to come to school, to develop solidarity and cooperation, to work not harder but better and to learn how to learn. We aspire to make students' work at college and home better and less of a source of conflict, while developing cooperation between peers and optimizing study hours which have become more numerous following the reform.

### Description

On the one hand, Louis Timbal College is engaged in an approach that should allow the students (and their parents) and the teachers to better understand their intellectual functioning in the framework of learning. On the other hand, so as to develop a sense of belonging, a policy of citizenship was introduced at the same time, supported by the creation of a Council of Student Life.

### Method of implementation

Proximity training offered four years ago to the pedagogic and educational team made them sensitive to self-esteem, brain functioning and yoga. *Positive Re-entry* began in September 2015. It is based on self-esteem and the development of methods for "learning how to learn". It takes place over several time periods and under various aspects:

1) *The day of re-entry*, 1 September 2016: for 6th graders, this involved preparing a personalized and caring welcome for the child and their family, creating a warm atmosphere and softening anxieties. This was achieved by workshops with exercises around self-esteem and the best knowledge of others and the establishment. "We welcome first of all human beings, and subsequently students".

2) For 9th graders, each of them is asked to explain their "vision" of success, performance and the goals which they may have set in their last year at this school, and they also have the chance to express the natural worries which may emerge at the moment of making a choice, to further strengthen cohesion and solidarity. The next day, an adapted version of work on the same thing is offered to the other levels.

3) *Workshops*: on 16 September 2016, beginning from the idea that everyone is different, 6th graders have a morning devoted to deepening work around methods of learning. Everyone learns differently, but everyone is most effective at work by using a method suited to their personal functioning.

4) *Learning profile*: on 26 and 27 September 2016, all 6th graders pass tests to establish, at a given time, their learning profile. This is not meant to "catalog" them, but rather to alert them to their strong points and their points to work on. By understanding ourselves better, we work better. Beyond the learning profile (auditory, visual or kinesthetic), these tests also concern the motivation and personality of each student. Teachers will thus have a profile of their class and may also embark upon personalization, creating partnerships between students sometimes with the same learning profiles and other times with complementary profiles.

5) *Exchanges with parents*: on 29 September 2016, the main teachers of 6th grade and the librarian teacher (in charge of the profile tests) met the parents of students to provide an overview of the start of the year, present and explain the results and give them their child's profile. At this meeting, parents were also invited to do these tests, so as to better understand their importance, as well as to better adapt the help given at home.

6) *Student diary:* the "homemade" student diary contains a guide made by teachers to suggest methods of working usable at home, mainly in the subject areas of French, math, English, etc. and transferable across all disciplines.

7) *New study*: 90% of our students use public transport in the afternoon, and have more study hours after the enactments of the school reforms. Given the significant amount of work not done at home, study hours were reorganized from September 2016 to set up collaborative and solidarity-based work. In this way, in association with and under the responsibility of the School Life team, several zones were created in a dedicated room. Each zone was marked out on furniture with a color code, and the overall layout was totally redesigned. Students can choose individual tables, double tables or blocks, which can accommodate up to four students. Several sofas can also accommodate students in a space intended for reading. Several green plants were also put in place and looked after by students. They were allowed to circulate freely between the different zones, depending on what they have to do. There was no longer an office for the AED (or a computer), which encouraged them to move around the different spaces. Mind-maps, multiplication tables, grammar rules, etc., requested or created by students, were displayed in this room. Study is therefore no longer silent but collaborative and enables a real struggle against inequalities. It also leads to "partnerships between students of different classes and levels". The role of the AED is thus different (and more rewarding), because in addition to simple requests, their mission now consists of supporting collaborative work and solidarity between students. They also give an overview for the students of the work to be done for this day and the next. They also have a workbook containing the profiles of the students and can thus suggest "associations" between students. They can also benefit from exchanges with teachers in following up those students in greater difficulty. A survey carried out in mid-October on this new organization among all the students and AEDs already shows positive results.

**Three resources or points of support**

– A particularly motivated and engaged group of teachers;

– the school reform and its structural implications have given the opportunity to offer new pathways such as *New Study*;

– personalized support benefits from work begun through the "learning college" approach.

**Difficulties encountered**

– The need to create progress reports with the teaching team all through the year;

– the majority of teaching staff dividing their work between two establishments;

– the positive attitude of parents is not necessarily accompanied by a greater commitment on their part to following up with their children.

**Resources mobilized**

The Teaching Council and the Council of Student Life have been consulted and the project has been validated by the Board of Directors. Financial aid was provided by the local social center for decoration and design of the study area (in particular for buying green plants). The CARDIE authorized overtime pay of 500 euros.

**Partnership and content of partnership**

The first partnership involves the parents of students who, through the local parents' council, financed the procurement of tests. Partnership with the CARDIE of the Limoges Academy for assistance with evaluation of this initiative in constant evolution for more than three years. The Limoges AROEVEN (regional association for educational work and holidays in National Education) for all their complementary work on citizenship, solidarity and the development of a sense of belonging around the Council of Student Life.

**Possible links with research**

*Les 7 profils d'apprentissage*, by Jean-François Michel, Eyrolles; *Travailler mieux pour bosser moins*, by Jacques Azam and Myriam Germain-Thiant, Editions de La Martinière jeunesse; *Apprendre autrement avec la pédagogie positive*, by Audrey Akoun and Isabelle Pailleau, Eyrolles; *L'apprentissage cognitif*, by Jacques Tardif (http://www.ac-grenoble.fr/occe26/printemps/tardif/cognitif.htm).

**Evaluation**

Evaluation/indicators:

– A questionnaire on *New Study* has already been offered to students in October. The operation will be renewed at the end of the school year. There has also been a questionnaire for families, on *New Study* and the *learning how to learn* approach.

– Parent–teacher meetings in the 6th grade are a chance to measure changes in work done outside of class.

– The Principal Education Adviser will ensure more precise statistical monitoring of work not done.

– Requests for guidance from students at the end of 9th grade concerning professional or general training, which is more demanding or more geographically remote, will be an indicator of the evolution of ambitions.

– The evaluation of the action, in particular its pedagogic impact, will be carried out by the Teaching Council and by class councils, with the goal of improvement in school results and in competences acquired. There remains the measure of the "well-being" of students and staff, which seems, scientifically, more difficult to grasp.

Documents:

– The project on video (slideshow), URL: https://www.youtube.com/watch?v = yuC4wy QBBqQ#action=share.

**Recorded impacts**

*On student achievement:* the first achievement relates to the fact that students come to school with pleasure. They say that they are happy and more confident in learning. The amount of homework not done (and the resulting punishments) is decreasing. New attitudes, showing more solidarity, have appeared in study. They have grasped the benefits of working together in *New Study* and the "partnerships" become natural and may spill over into class. These partnerships have moreover formed between students from different classes and different levels. The atmosphere in 6th-grade classes is particularly positive, with regard to the "difficulties announced" during the 5th–6th grade liaison meetings of last July.

*On teacher practices:* the main 6th-grade teachers are completely engaged in the process with other colleagues drawing benefit from this "philosophy" in teaching and in their relations with students. The adaptation of teaching is more efficient. This "philosophy" shared between teachers also extends to interdisciplinarity and personalized support.

*On leadership and professional relationships:* expanded relationships between AEDs and teachers, richer and more frequent pedagogic and didactic exchanges between teachers: for a head of school, this work of reflection and change in practices is particularly stimulating. It allows us to place the human being, rather than the student, at the center of everything. It is obviously a question of personality, but this approach allows staff to come together and sincerely express their expectations and difficulties. It is another way of being open, better recognizing each other and concretely sharing a school project as more than a piece of institutional writing. It is a challenge, because it remains a challenge to bring people together.

*On the school/institution:* the climate at school is positive, students and teachers are known to each other, there is no absenteeism and very little indiscipline.

More generally, *on the environment*, the college now has a more positive image. It has recently obtained three certifications (boy–girl equality in sports, Agenda 21 and Sustainable Development in Education). Work on communication is important. The Internet site receives around 2,000 visits a month. Exchanges with other schools in the sector are rich, and regular relationships with town halls and other institutions are positive. The school is also open to media and often available for new experiments. The school often receives inquiries about the functioning of the CSL.

**Box 2.4.** *Extract from a written account of innovation in a junior high school*

This account, taken completely at random, shows a clear evolution in the concept that instigators have of such promotion. In fact, if 30 years ago promotion of innovation by institutional leaders resembled a dry summary, we can see in the present account a narrative, mixed with very different literary genres, including in terms of content. Research concerning the literary genre of a written promotion of an innovation would be doubtless instructive and allow us to measure the impact of reading and consultation of videos beyond simply counting Internet hits.

In addition to this question of the choice of which innovations to promote, the problem remains that of the method of communication. Is written communication the most relevant way to communicate a practice? Might there be other forms of communication? Is school not already too weighed down by writing, or does this latter not enforce a singular approach to the development of learning, including for innovators?

Might we consider promotion of innovation without written media?

Comparison of the three accounts given in Boxes 2.2, 2.3 and 2.4 reveals an evolution in the form of recounting and promotion of innovation that we find on the ministry's Expérithèque Website.

Writing, reflecting its role as the major medium of schools, has in fact become the most obvious form of communication for leaders with a dual objective: writing about one's practice is an activity considered to be formative for those who write[25], and this written account is easily communicable as a familiar vector for all teachers.

Thus, the idea has emerged of making innovators write about their practice, in a common format imposed by the administration.

This idea is supported by a current of thought of analysis of professional practice as a means of training. Briefly, making innovators write is said to allow promotion of the innovation and development of the competences of these professionals[26]. For some years, as part of continuing education, innovators who agree to write about their innovation have benefited from hours of leave and from support provided by the academic leaders of innovation. Publications have appeared based on the themes, which these innovations deal with.

We have carried out a study concerning the diffusion of innovations, in particular through writing[27]. The original idea was that, certainly, we can provide information about innovation through writing, even using readymade forms, but the real objective is to invite the reader to modify their own practice. This is where the difficulty lies: we can invent any kind of medium that we want, but if we have not thought about the effects which these produce on the recipient, we will never know if this has had an impact on them or, even less, inspired them to change their professional practice.

25 CROS F., *Écrire sur sa pratique pour développer des compétences professionnelles,* L'Harmattan, Paris, 2008.

26 BLANCHOT M., *L'espace littéraire*, Gallimard, Paris, 1955. See p. 108: "Writing changes us. We do not write according to what we are: we are according to what we write".

27 CROS F., "La diffusion des innovations comme expériences", in MORISSE M., LAFORTUNE L., CROS F., *Se professionnaliser par l'écriture. Quels accompagnements ?,* Presses Universitaires du Québec, Montreal, 2011.

How to successfully invite the recipient of a message to move towards modifying what they do? It is in this sense that we have considered innovation as an experience and, as such, not reproducible. Therefore, if experience is neither reproducible nor really communicable, what is the point of making people write by setting these goals for them?

The affective experience of carrying out an innovation is strong, as it often involves risk-taking, leaping into the unknown. This dimension is little mentioned: in the end, is not innovation above all else an experience fraught with emotions, aspirations, joy, anger, love, force and respect? The activities carried out are only the outer layer of the web of emotions involved in innovation. In this case, can we communicate this emotion in action in a standardized form?

This is doubtless the issue in diffusion of innovations for the purpose of transforming practice. This is why we wanted to test a hypothesis regarding writing and the transmission of these emotions. Emotions may be the basis for learning and the locus of modification of practices[28].

If we consider innovation as an experience for those who have implemented it, what would be the use of the experience for someone else to develop themselves? Is it not common sense that experience cannot be transmitted, only felt?

Maybe it is not as categorical as this, and we may suppose that something is diffused, depending on the proximity of the reader with regard to the written experience, through similarities, understandings, etc. This active emotional experience would provoke reverie and would have a dreamlike appearance. We must dream to innovate, to aspire to an unreal better and to develop our imaginary and our imagination.

We thus began from two *a priori* concepts:

– On the one hand, innovation, in its multiple meanings, as it is considered and used in education, is experienced by innovating subjects as an experience in that it contains a strong emotion, which is marked in the memory of the innovator. And, in this sense, it is for the latter a memory which gives them the strong will to encourage other colleagues to engage in this pathway.

---

28 DAMASIO A.R., *L'erreur de Descartes: la raison dans l'émotion*, Odile Jacob, Paris, 1995.

– On the other hand, the innovator wishes to not only make their innovation known but also export it so that others may share it, even if this attitude is not without ambivalence[29].

If we remain with the practice of writing as a means of diffusion of innovation, what should we do? What type of writing should we ask for? Passing from experience as we have defined it to writing about one's experience means a change of register: it is moving from a semantic of action to a semantic of intelligibility of an action! In other words, the formalization, the putting into words of an innovation, is not the innovation itself, but a linguistic reconstruction according to the constraints of a particular syntax with a common vocabulary (intertextuality).

In aiming to mobilize readers, it is possible to call this writing performative[30]. This kind of writing has hardly ever been studied by education researchers. What are its ingredients? Is it a particular discursive genre which not only leads to the enactment of what the writer describes, but performs the act of inviting the reader to act? Is it an argumentative form of writing with a rhetoric of persuasion? Is it a writing that constitutes the author or their activity as an object of desire?

This is where we see the process known as typification, which allows us to understand what is produced when moving from an innovative experience to writing about this experience, which is considered meaningful by its author (as they describe it and we do not describe all our experiences, but only the most significant ones). As Schutz writes[31], "meaning is not a quality inherent in certain interpretations which emerge into our stream of thought, but the result of interpretation of a past experience which we reflexively envisage from the viewpoint of the present moment". The contents of writing about innovation thus moves away from simple testimony (often unfaithful, but to what?) or the description of a perception, to enter the world of meaning.

29 We have seen innovators who, once they have realized that their colleagues have adopted the same innovative practice, have complained of theft or even of the distortion of their innovation.

30 AUSTIN J.L., *Quand dire c'est faire*, Le Seuil, Paris, 1991.

31 SCHUTZ A. *Le chercheur et le quotidien: phénoménologie des sciences sociales,* Meridiens Klincksieck, Paris, 1987.

The teachers we have met who have read these innovation sheets indicate that reading these sheets did little to incite them to change their practices, dropping everything else, but sometimes gave them ideas. What they wanted was to go and see in these schools how these innovations were carried out, to become immersed in the innovative activity itself[32].

It is indeed possible to visit innovative activities, but that can only happen at local level and in free time, which reduces the range of variety of innovations implemented all over French territory[33].

These teachers add that exchanges like those of the national journal on innovation, organized annually by the ministry, or inter-academy meetings, give them more interest in innovating: they had had oral discussions with innovators and had been able to raise the questions which were important to them. Rather than discussing innovations, would it not be more relevant for the innovators to have discussions?

But this is no reason to reject writing as a mode of diffusion of innovation; we only think that we have not yet found the type of writing which may surpass the level of information to invite the reader to action, even if this action is not a reproduction of what is described.

In this sense, we thought that writing clearly displayed as related to fiction was a perspective to try out in the aim of diffusion. The idea is that this type of writing invites the reader to view the experience of the innovator as relevant to the experience of the reader, giving an impression of verisimilitude.

One of the strong arguments of the fiction specialist Schaeffer[34] consists of raising "the unity of every person's representational structures, whatever the usage. And the fictional nature of a text is among the contextual markers which indicate the genre of discourse with which the reader is dealing and, therefore, the correct cognitive attitude for the reception of the text" (p. 27).

32 As of now, there exist, often supported by Canopé, videos on YouTube, but given the very short duration of this type of media, this brings little to the spectator even if it strengthens the identity of the school and the innovators.
33 On this subject, see the section on "mirror neurons" above (footnote 18).
34 SCHAEFFER J.M., *Pourquoi la fiction?* Jacqueline Chambon, Paris, 1999.

In other words, writing about innovation using fictional techniques would evoke experiences of the same nature in the reader. Mimetic fictional production is one of the distinctive traits of human beings, a sort of "lived world" (Habermas, 1987)[35], "that is to say, a space of intercomprehension in which people move and which provides them with possibility of mutual understanding in interactions with people" (p. 216).

Fictional writing about innovation would put the reader in a position that invites them to modify their field of representation in the professional world. It would allow them to suspend the real consequences in life (one may at the same time enter into the narrated action and leave it without any damage): the reader understands the risks experienced by the innovator without experiencing them, except through fictional interposition. Regardless of whether what is told is faithful to what actually happened, the writer speaks of their experience to another colleague in the hope of inviting them to share concretely, in the action, the innovation and to place it in their active intentions.

According to King (2000)[36], a "description begins in the imagination of the writer but must be completed in that of the reader" (p. 207). We are dealing here with verisimilitude. To invite another to change their behavior through reading, the most important thing seems to be not truth, but belief. This collection of fictional representations is constructed as easily (if not more so) as a collection of factual assertions in a standardized form and might achieve its goal.

It is enough for the reader to believe without seeking to fool them: the writing appears plausible. It is not by the writer exhausting their memories, getting as close as possible to the facts, that they influence the reader more! The words used are only reflections of the real, what is called a typification. For example, Magritte's famous "*Ceci n'est pas une pipe*" (This is not a pipe) reminds us that the painting is a representation of a representation, which does not stop us seeing a pipe in it!

35 HABERMAS J., *Theory of Communicative Action*, vol. 2, Fayard, Paris, 1987.
36 KING S., *L'écriture: mémoires d'un métier*, Albin Michel, Paris, 2000.

In other words, we hypothesized that, rather than producing filled-out fact sheets in a more or less standardized manner, we would invite innovators to produce a narrative text, to somehow set the scene, to "introduce the plot" (Ricœur, 1984)[37]. The innovator's writing does not, however, need to go in the direction of presenting an heroic epic.

We tested this hypothesis by forming three groups of six innovators each at primary school level, in three academies (X, Y and Z). These innovators were identified through CARDIE in three academies based in different environments: rural school, urban school and semi-urban school (two innovators per academy). The innovation in itself was of less interest than the fact that the innovators designated themselves as such and that the academic system supported them as such.

The first two groups were invited, one to write according to the standardized form (with categories already mentioned such as: context, objectives, procedure, evaluation) and the other group was invited to produce fictional writing for which we had prepared them. The distribution was made according to Table 2.3.

| Group | Academy | School | Type of writing | Number |
|---|---|---|---|---|
| G1 | X | Rural school | Standardized form | 2 |
| | Y | Urban school | | 2 |
| | Z | Semi-urban school | | 2 |
| G2 | X | Rural school | Fictional writing | 2 |
| | Y | Urban school | | 2 |
| | Z | Semi-urban school | | 2 |
| G3 | X | Rural school | Reading the two contrasting writings | 2 |
| | Y | Urban school | | 2 |
| | Z | Semi-urban school | | 2 |

**Table 2.3.** *Distribution of populations according to types of schools and academies*

The first group (G1), comprising six innovators (self-described as such and willing to write about their innovation) belonging to three different academies and schools in geographical terms, was invited to produce a

37 RICŒUR P., *Temps et récit*, vol. 2, Le Seuil, Paris, 1984. For Ricœur, "introducing the plot" is "an integrative dynamism which creates a single, complete history from various incidents, in other words transforming these incidents into a single and complete history".

written account according to the official framework shown on the Board of Education's site, while still having the possibility of describing their innovative action in the body of the form, using the space for execution.

The other group (G2), comprising six innovators (of the same type as the preceding group), was invited to write about their innovation in the form of a story, where there was a beginning and an end with a plot, a narration which presented characters with their emotions, feelings, reactions or even rejections[38].

For the reader to understand the difference between the two types of writing, we give in Boxes 2.5 and 2.6 an example of standardized writing and an example of fictional writing (these are only short extracts).

*Title of the innovation:* **"Appropriation of three-dimensional space"**

***Context:*** this concerns children coming from disadvantaged socio-cultural backgrounds and experiencing difficulties at nursery school. The setup of the games room only allowed limited exploitation of space. The teaching team was motivated by this finding.

***Goals:*** to address deficiencies relating to background and environment, to offer more complex and more varied situations, to take account of the fundamental role of the body in the overall development of the child (motor, affective, cognitive, social), to promote a better spatial structure, in particular using the third dimension, and to make parents aware of their children's needs in the motor domain. As for the teachers, it involved permanently preserving a collective reflection, enlarging and enriching pedagogic practices and generating consistent research into training.

***Procedure:*** to design the richest possible structure, the teaching team called on external stakeholders. The teachers gave themselves consultation time to develop teaching sheets based on the capacities implemented or based on the material offered, so as to set up varied and different situations, design grids for observation and evaluation, and analyze them and adapt the situations in order to ensure continuity in gymnastic practices from three-year-olds to five-year-olds, etc.

**Box 2.5.** *The standardized form*

38 Precise instructions were given with support so as to make it clear to the writers that what we were looking for was not the most faithful account of what happened, but for the writer to retrace their emotions and describe what motivated them in their innovative action.

"I have been a teacher in a nursery school in a working-class neighborhood for eight years. The school, situated right under the windows of the social housing block where the families live, is tiny: the children feel stifled, that they are constantly being watched by their big brothers. The children sometimes come very early to school and wait there for the doors to open, because their parents work in the city at unsociable hours.

Between colleagues, we have often talked about this situation, which is becoming more and more intolerable for us. One day, we decided to react and to find activities where the children would not encounter the smallness of their apartment, but a rich, colorful environment with new people. So, we invited clowns – not the kind from the circus, but the kind who invite children to play with them – to portray characters from legends or fairy tales. We also invited children's authors to come with their books and images. Finally, a real paradise for these children: their eyes lit up and three weeks later, the four students in the class about whom we were most pessimistic came up with colorful crayon drawings, which they had discussed with their friends, etc."

**Box 2.6.** *Fictional writing*

Next, once all the texts had been drafted, after a month, we gave these two types of writing to six teachers from other schools to read, teachers who were not designated as innovators, but belonging to the same academies. We waited another month and then carried out interviews with the readers of the two types of writing to identify the effects of this reading. Five of the six subjects mentioned, in a significant manner (that is, giving concrete examples of how they had transformed their practice inspired by this reading), the innovations described by using a story introducing characters with emotions, struggles and doubts.

Obviously, these were self-descriptions, but it would not be impossible to move on to a classroom observation before/after reading to identify changes in the readers' practices and to try to bring them closer to the innovations described in the fictional manner.

Such an inquiry would no doubt require a stronger methodological apparatus with a more significant sample, something we could not provide in the context of this period.

However, this little bit of research does show that when the reader has access to their own subjectivity through encountering that of the writer, and through a mediation of appearance and fiction, the impact on their desire to innovate is stronger than from reading a written account in a more administrative format.

Yet it is not a matter of asking the innovator to write stories or a novel but to play with the methods of fiction through a narration where verisimilitude draws in the reader in their own subjectivity. Such writing is often rejected on both sides as not showing the professional character required of all teachers, including in explaining their work.

Such writing requires support from innovators, because fictional writing, especially when destined for several readers, cannot be improvised and might be part of the ongoing training of teachers, looking for emotion and affect in a profession where intersubjectivity is foregrounded.

We raise the question of the type of writing that the institution generally requires of its teachers, which is more in the style of accountability rather than indicating ways of professional learning, which follow the nonlinear path of the adventure involved in any innovation. Now, the line between self-referential social objects (which is often denied) and those referring to a reality dependent on the representation that each one makes of them is fine, if not non-existent.

Such an interrogation leads us to rethink the training of teachers through reading the writing of innovators, and this has two aspects: on the one hand, in the very process of creating written accounts of professional activities and, on the other hand, in the ignorance which remains about what works for possible change in teachers' attitudes, that is, the force of the fictional in training itself.

### 2.5.3. *The perpetuation of an innovation within an educational establishment*

We conducted a study in the psychosociology of education on the question of the permanent establishment of innovations within an educational institution, trying to respond to the following question: what causes an innovation to endure within an establishment, or become a routine, while another, almost identical innovation, disappears very quickly? What parameters are at work[39]?

This question emerged following a remark by Antoine Prost, who said in 1997: "we can even consider that all examples of innovation are failures to the extent that their generalization is not successful". Do we then conclude that all innovation is local and cannot be separated from its place of birth? Are there no constants that would allow us to understand the innovative process? If we know that an innovation is not transferable like an object which is decontextualized to be recontextualized elsewhere, maybe we could envisage an identical process, a kind of model of intelligibility of the establishment of an innovation, a communicable model?

The problem of transferability already discussed shows that a practice cannot be transferred, but that it can give birth to new practices through the spread of ideas or know-how. The question here is rather that of the melting pot in which an innovation develops and is sustainably established. What would be the conditions for an innovation to become established and develop within an organization?

Our research team followed six junior high schools, contrasting in their ways of recruiting students and geographic location, in which the innovation was almost identical (we intentionally chose an innovation requested by the ministry and mentioned in the national innovation program. This was "diverse learning paths").

From observations, interviews and a three-year follow-up, we observed that what made the difference between a school where innovation becomes established and perpetuated with the agreement of all and a school where innovation is abandoned, either because of the departure of its bearer, through weariness, or through non-recognition, was based on the following seven conditions[40], in a way verifying Callon's theory of translation:

– the existence of what we might call a "laboratory of ideas", that is, the presence of places in the school where teachers and staff can initiate

39 Cros F., "Émergence et installation de l'innovation scolaire: pertinence de la théorie de la 'traduction'", in *Transformer l'école*, De Boeck, Brussels, 2004.
40 Ibid.

proposals, sometimes unrealistic, sometimes utopian, but which are expressed and heard;

– the fact that some of these ideas must be put into form, that is, written and published, and that they have a place at working meetings, in discussions, on ordinary days, etc.;

– the taking up of these ideas, whether by the head of school, by linking them to official texts or by integrating them into the proposals of the school's project;

– mobility of staff within and outside of the school. For example, if a teacher over a 30-year career has only ever taught at one level, that is prejudicial to the development of innovation. Innovation needs flexibility in structure and responsibility;

– promotion by the head of school, an important level for transformation of the school, of the existence of networks based on profession and not by affinity (particularly because an innovation often brings together people who have a certain affinity, which causes problems for the viability and even the sustainability of the innovation);

– the existence in the school not only of moments of regulation, but also of places where disagreements or controversies are expressed and produce recognized negotiations;

– finally, involvement in innovation of all the institution's staff, whether to critique it, enact it or develop it.

In other words, regardless of the structure of the schools, these elements appear to be the necessary conditions for the establishment of innovations in educational establishments.

## 2.6. Critical analysis of examples of innovation in schools

Access to innovation is not an abstraction; this is shown in the very achievements of innovators. We have established over our many years of interest in innovation that the most valid approach[41] is to go to the place of

41 By "valid" we mean the idea that innovation as a new action, as risk-taking and uncertainty live from day to day, and that an *a posteriori* view does not allow us to understand the "creativity of this action", as Hans Joas notes.

production of innovation, during the process, a journey which is chaotic, uncertain and filled with hope.

In brief, the field speaks to you, assuming that you do not just spend a single-half day there, but return, follow its developments, fervor and disappointments. This is difficult, and it requires a high degree of availability, quite aside from the fact that you must get the innovators to agree to such immersion.

Access to innovation is therefore often provided through a presentation of the innovation by its initiator or their direct manager (or team), which enthusiastically explains what has been done (or what they think they have done) or what they are doing.

And we thus see an incredible variety of innovations, of different magnitudes, at all levels of school. However, we must admit that the ministerial procedure, which is in place (we will discuss this in more detail later), channels these innovations so as only to keep those which correspond to the national priorities for development of the educational system (for example, introduction of ICT into teaching, development of citizenship, concern for gender equality, etc.).

Let us add that, independently of this channeling, the French educational system, like any porous professional environment, is under multiple influences, the effects of fashion or of interpretation of the latest discoveries from research, etc. It is in this way that we see development in the activities of groups of students, overtures towards a number of partnerships, multidisciplinarity, interventions from students' parents, pedagogical sequences where doing is the first priority[42], the central role of emotion, multi-age classrooms, inclusive classrooms, the entrance of spirituality, work on the body, etc.

The framework is there for a reason, and the discourse of inspectors tries to demonstrate that innovation is the result of "necessity"[43], particularly in

42 On this subject, see BERLINGUER L., *Réinventer l'école. Une école de qualité pour tous et pour chacun,* Fabert, Paris, 2017.

43 Speech from the dean of inspection of Agricultural Education, 30 March 2017.

taking into account differences in the learning profiles of students. The areas in which teachers are invited to innovate are the following:

– diversification of practices;

– digital technology;

– promotion of sequences in the professional or social environment;

– evaluation;

– pedagogical steering (synergy between management and teaching teams).

Looking at many innovations, we can see at the same time common concerns and similar responses. In other words, schools are moving in the same pedagogical direction, but at different rates.

To those who say that schools are behind in the field of innovation, we believe that this is not correct. School is moving, granted in an uneven manner, depending on the teaching teams and actors in each school. The problem is not, as claimed, with the teachers, but with the social context of which they are part.

As we said in the Introduction, society moves by leaps and uneven advances: some people are surrounded by software technology, while others refuse such an intrusion. For example, the invitation to work in interdisciplinarity is not obvious in high school classes whose content remains in disciplinary silos.

An innovating teacher is often accused of having converted to a radical spirit of acceptance in order to get out of the work which they were thought to be contracted to do. "Basically, I am a PE teacher, but it is obvious that I've been led to do many other things"[44]. This testimony shows that innovation requires stepping out of professional frameworks as they are usually defined and getting involved in tasks for which teachers are not always trained and prepared. It is as if a butcher was asked to help their customers, beyond providing excellent meat, by composing their menu and

44 A teacher from the team of an Agricultural Education professional high school, at the UNREP (National Rural Union for Education and Promotion) "Training to Succeed" day, 30 March 2017.

organizing their meals for the week! Where does the teaching profession begin and end?

Many associations governed by the law of 1901[45] strive to develop the joy and happiness of children. We do not believe that public school would be wrong to have the same mission. The question is simply that of connecting overall educational concepts and learning in the goal of national cultural development, where acquisitions which are not "natural" are mandatory. For example, discovering the classics of French literature does not arise from a child's spontaneous request, but from the strict contribution of the teacher. Everything is done in such a way as to convey messages.

Remember, about 40 years ago, the quarrel which arose between two educationalists whose visions of pedagogy were very different: the first thought that acquisitions and knowledge were necessary to make a child into a citizen of French culture;[46] the second opted for a differentiated pedagogy, which involved all the activities likely to develop the child over a preferably and initially educational field, without overly worrying about building content considered by the nation to be indispensable for all citizens. According to the times, the balance dips in one direction or the other.

There are no winners and losers, because we believe that designating some people as "resisting" innovation would imply that these teachers intentionally refuse to innovate. Now, that seems complicated, and if some teachers refuse to change their practices, that is not because of some wicked strategy of blind systematic opposition, but from conviction.

In fact, school is implicated in society's contradictions and it is normal to see different positions concerning certain innovative activities. The most important lever with which to move such people is persuasion, and involvement in and acceptance of a new idea considered to be more beneficial for both the teacher and students.

---

45 The law of 1 July 1901 concerns the freedom of creating associations based on established principles.

46 Remember the conception that Alain had of learning in his book *Propos sur l'éducation*, where he wrote: "You say that we must know the child to teach them, but this is not true at all; I would rather say that you must teach the child to know them".

Often, resistance is based on a lack of problematization. In fact, the innovator does not innovate for their own sake, but to improve a situation which they find negative. They have problematized the situation in such a way that they feel the need to find a solution. A better way may be built on a problem that is discovered. For example, failure at school has become a major problem and attitudes concerning this difficulty allow us to identify the explanations given for a desire to innovate, or on the contrary, an absence of desire to innovate.

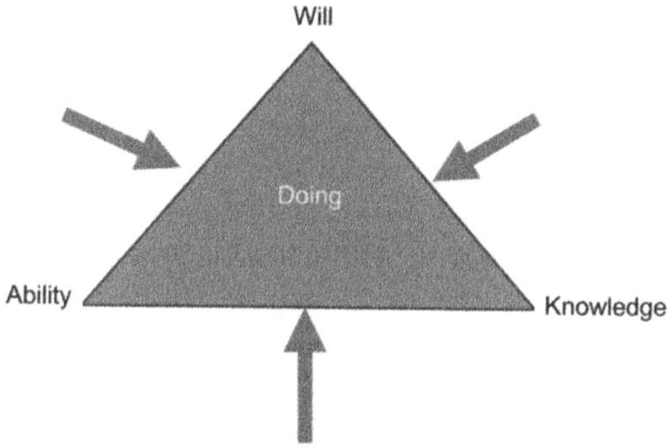


**Figure 2.2.** *Active principles in reasons to innovate*

If all the teachers agree on this concept, problematization means discovering possible actions to overcome this difficulty. Three kinds of opposing reactions may be observed:

– the "knowledge": refusal is linked to a lack of knowledge, understanding or know-how, and actors in this category require conditions and clarifications to overcome this difficulty;

– the "will": there are many actions which will lessen this difficulty, but administrative or legislative obstacles will be invoked;

– the "ability": there may be all the will in the world to surmount this difficulty, but a lack of resources or support will be invoked.

Every person wishing to innovate in response to a deficiency, unless they verify whether everyone recognizes this deficiency, is advised to anticipate the consequences of the possible enactment of an innovation in terms of "knowledge", "will" and "ability", so as to analyze possible resisters and work with them to progressively move forward toward a moderated and negotiated innovation.

Development of innovations begins, in most cases, with a time-consuming period. In fact, setting up an innovation requires investing time (sometimes money) and energy. Not everyone is ready to go for it. In the 1970s, we carried out a study on the establishment of teaching teams in the experimental junior high schools followed by the INRP[47] and we observed that many colleagues refused to participate in these meetings under the pretext of not having enough time[48]. However, the teachers who overcame this apprehension and attended these meetings found benefit in them, certainly late, but significant.

In fact, teachers who are members of a pedagogic team have been able to build upon their interventions with students, harmonize student performance evaluations and share knowledge on pedagogic practices with each other, to the point that if some teachers are absent (because of sickness or training), they can be easily replaced by their colleagues who temporarily take over their classes. Similarly, the construction of teaching sheets and didactic tools is found to be simplified by these mutual exchanges. However, this gain could not have been seen except after quite a long period of time, which some were not willing to tolerate. Innovation won, but then, it often wins!

Here, we will present you with some innovations, some which have already been mentioned, in their content[49]: at elementary, junior high and high school levels, in such a way as to lead a critical and interrogative analysis of these activities in the choice of their object of innovation (what is considered "new") and in the reflections which they generate. They are drawn equally from general, technical and professional streams of National Education as well as from Agricultural Education.

47 INRP: National Institute of Pedagogic Research, replaced in 2011 by the French Institute of Education, located at Lyon and hosted by the *École normale supérieure* of Lyon.

48 We must add that this observation showed us that teachers are very talkative and tend to spread out, and that training in collective and team work is indispensable. The agenda was rarely respected.

49 Our objective here is to identify the origins of the inspiration for these innovations and the exploitation of this inspiration. We will not therefore be discussing the enactment of the innovation, but what constitutes it as a new object in its use for learning.

### 2.6.1. *Elementary level*

In the Montpelier academy, a school innovated in cycle 2 by setting up correspondence between a 1st-grade class and a class in a New York school at the same level, and a 2nd-grade class with a New York class at the same level. The start of English teaching recommended in the official texts for elementary classes allowed them to engage in this continuing correspondence using different media (postcards, letters, videos, art projects, poems, songs, etc.). This correspondence went through a secure Internet site managed by the educational department of the French Embassy in New York.

### 2.6.2. *Junior high level*

We return to the innovation by Timbal College in Chateauponsac, where the primary goal was to develop the well-being of actors in the college through a series of new activities such as help with homework at a distance, the mutualization of this help and the Council of Student Life along the same lines as in senior high schools working on many cultural activities leading to feeling good within the establishment. This set of activities produced well-being, sociability and mutual respect.

The initiators of these activities were teachers who had had the idea to give students a speaking space in a supportive and democratic spirit. The first days of the year are important, particularly for new students. From the range of pedagogic tools, the teachers drew the concept of multiple intelligences in the hope of helping students choose between approaches, which might be more auditory, visual or kinesthetic. Staff named their school a “learning college”, doubtless showing that this set of activities leads to better learning by all, and thus by the school as a personified entity.

### 2.6.3. *High school level*

In a professional high school in the South of France, a preparation class for the agricultural CAP (CAPA) in personal services involves 250 students over two sites. Teachers receive continuing education in project pedagogy.

Three teachers are involved alongside supportive school management; they have also benefited from the renovation of the CAP. These teachers are supported by a social science research laboratory.

The three teachers first studied what the professional practices of their future students would be when employed. They then created between them an “ideas box”, where suggestions for new activities could be deposited on bits of paper, by students and teachers alike. The box was checked every week, and ideas were considered for their feasibility and organization.

The innovation concerned a project-based pedagogy, which was strongly oriented to the professional and social world, as in the project entitled “Enterprise for Learning” in view of creating a mini-business, or organizing story-reading in a nursery school or developing a set of entertainments for elderly people living in care homes.

What do these three examples suggest?

1) So-called “new” innovations resemble many previous or contemporary innovations.

2) Innovations are open to interpretations, concepts and representations arising from the results of research.

3) Innovations are only reflections of new pedagogy or old pedagogues, whose diffusion seems to occur centuries after their death, with no mention of their input.

### 2.6.4. *Innovation never comes alone*

The three innovations examined, like many others, are like the tree that hides the forest. In fact, in pointing out these innovations in the public education system, we are struck by their identical appearance in other schools, without there being any real possibility of the networking of these very similar innovations, as if they were so different and singular that it would be impossible to bring them together.

In fact, innovations throughout the territory are very similar while believing themselves unique, in a series of successive waves. Thus, correspondence between French and other schools is not the prerogative of a single school. Even if we say that all innovations are unique, it does not prevent the spirit of the times from delivering quite similar ideas. We find the same thing in junior high, with the development of cultural activities giving students a sense of well-being favorable to learning. Finally, the high school, with the intention of giving direction to teaching, carries out projects which may develop all of the competences necessary for learning. Other high schools also carry out this kind of activity, whether they call it innovation or not.

### 2.6.5. *Innovation is not born of nothing*

The three innovations cited relate to ideas circulating in the professional environment, whether instilled by the trainers themselves or shared at meetings between teachers, without any guarantee of certainty and often in a simplified manner. The example of Gardner's multiple intelligences, following Antoine de la Garanderie, underlining that intelligence comes in many forms, leads to students confirming behavior that they thought was theirs alone (a self-fulfilling prophecy), through tests, for example, as in the Explorama software.

The video from Timbal College and the account of the agricultural high school attribute intelligence profiles to students, according to tests. Students, fond of this kind of categorization, confirm their dominant form of intelligence. If Gardner, and before him La Garanderie, have had the merit of questioning the relevance of measuring IQ, it is nonetheless true that the reified usage of their different types of intelligence, in assessing the student's learning process, is not necessarily the most appropriate way to ensure adequate and efficient pedagogy. In fact, the construction of a typology of intellectual profiles from which a pedagogy arises is more random: how do we go from a profile to a pedagogy which specifically responds to it?

The multiple intelligences approach certainly has the merit that it does not evaluate students according to a single concept of intelligence based on logic, rationality and conceptualization. And that has a double benefit for teachers: on the one hand, avoiding ranking students among themselves based on a single criterion of intelligence and, on the other hand, opening school to disciplinary content and pedagogical approaches, which are open and multidisciplinary. It loosens the grip of the dominant disciplines and allows us to put in place pedagogic activities, which are more fun and pleasurable.

The central problem arises from a possible ideological recuperation, which leads us to think that everyone is intelligent, in different ways. This consideration rests on aspects that are more humanist and moral than scientific. In fact, studies in differential psychology, beginning in particular with Thurstone, show that multiple intelligence profiles are really components of a single intelligence that is expressed through a G factor, called "general intelligence". In other words, intelligence resides in the capacity to organize these different components or profiles in a relevant way, and not to decree that students belong to such or such a component area.

In other words, many innovations emerge from the circulation of information, which is more or less well digested and whose meaning is collectively rewritten. The role of research is doubtless one of the ways of controlling certain deviations, on the condition that its results are well understood and not used to put simple and categorical conceptions on the market of teacher opinion. However, the borrowings (sometimes diverted) made by innovators attest to their thirst for change in their practices, without subsequently finding any support or aid.

### 2.6.6. *Is innovation only old pedagogy in a new wrapper?*

The three innovations examined show that when it comes to pedagogy, based essentially on the pedagogic relationship, there is nothing fundamentally new, as we emphasized when defining the concept. If we take each innovation:

– in the elementary school, if correspondence between classes in France and the United States shows the use of modern technologies, due to the development of virtual communication techniques, we cannot help but think of Freinet's pedagogy, where correspondence gives meaning to learning in the field of language, communication and articulation with socially significant actions. Free expression through poems, drawings, creating a class newspaper which is widely distributed, etc., is one of the elements of this pedagogy. What has changed from Freinet is the technological sophistication of the media allowing wider diffusion;

– in the junior high, the multiple intelligences approach recalls the approach of Maria Montessori who, from observing students, pointed out the different preferences in their orientation to domains such as the technical and the logico-mathematical, without exactly categorizing them in Gardner's way. Piaget, whom Gardner credits as a partial inspiration, had already distinguished more abstract forms of intelligence from the more concrete (preference for manipulating objects). The worry of this kind of approach is that "good students", or at least those who seem to have greater intellectual agility, use all these different forms of intelligence in an organized manner; the problem is their relevant organization and use. In other words, Gardner's conceptualization was latent in Piaget and Montessori, among others;

– in the high school, the project-based pedagogy on which the innovation is based cannot help but recall the pedagogy promoted by John Dewey (1938), involving actively engaging students in school tasks, which most often had relevance in the greater world. This is a way of developing the motivation of students. We also see here that innovation involves nothing new except that the social context has changed. Note that the project process was established very slowly within the school. In the 1930s, John Dewey launched his famous *learning by doing*, inviting educators to motivate young people by putting projects into action. It was in 1973 that, in France, the Fontantet ministry spoke of 10% of teaching time spent on work on common projects, then in 1979, the PACTE (Project for Cultural, Technical and Educational Activity) under the Beullac ministry, then the PAE (Project of Educational Action) in 1981 and the school project under the Jospin ministry in 1988.

This increase in power of the project approach occurred from 1970 to 2017, in more than 40 years of progressive introduction into the French educational system! Can we say that an innovation took a slow route from its founder (Dewey) some 30 years before, then a timid advance over 40 years?

The works of Paul Ricœur[50] praise forgetfulness in tandem with a promise of creating a blank slate to leave room for the new. Thus, in contrast to what one might think, innovation in education is based on the forgetting of previous innovations, the better to put into practice a novelty which is relative, but still experienced as such.

50 RICŒUR P., *La mémoire, l'histoire, l'oubli*, Le Seuil, Paris, 2003.

## 2.7. Innovation in schools and its support

In this chapter, we will not discuss the role that research, in its different forms, can bring to support for innovators. That subject will be dealt with in the chapter on the link between research and innovation.

The polysemy of the term "support", which is truly nebulous, includes several forms of *soft* intervention on others. Since 1990, as M. Paul[51] emphasizes, support has come to substitute for the professional and personal development activities, both formal and informal, known as guidance, tutoring, mentoring or coaching. These words are equally recent in the language of training, but they have been more and more abandoned in favor of the word "support", which seems to have a broader meaning.

The word support suggests a social relationship between people at the same level, somewhat erasing hierarchy or position of authority. It seems to imply a two-way relationship, but more and more frequently it has been used in group or collective situations or even on the virtual plane (we speak of support in relationship to setting up MOOCs).

The diversity of support practices, including when it comes to innovators, does not allow us to organize this fluid set in a clear manner. Yet, it seems that support is related to the contemporary image of training. Before, it was frequently thought that people were trained for such or such an activity, but now, the accent is on the persons training *themselves*, and thus the result of training depends on their desire, respect for their autonomy, their feelings and the confidence they have been able to build.

Support, including in this form, seems to better correspond to the help which an innovator or a team of innovators might need, when in uncertainty and doubt. Support would thus respond to their needs in a somewhat maieutic[52] way, with multifaceted support in a process of transformation of practices.

---

51 PAUL M., *L'accompagnement: une posture professionnelle spécifique?*, L'Harmattan, Paris, 2004.

52 M. Paul identified three forms of support: initiatic, maieutic and therapeutic.

Such support cannot come about without willingness on both sides, mutual confidence, and disinterest on the part of the supporter, if they are to help their supportee in their quest for understanding and modification.

The question to be answered is that of the objectives of this support and its institutional backup. If support can in fact occur between peer teachers, where a more experienced one comes to the aid of a less-expert colleague not by providing advice to follow, but in allowing them to formulate their own questions and find their own response therein, it is not always the same thing.

It is in this sense that we speak of supporting reforms (and this is no longer at an individual but an institutional level, to help them "swallow the pill" of a reform decided at political level, linking the word in its general sense and reality on the ground). This is a sort of diversion of the original sense of support in favor of an institutional framework in which pedagogic advisers and inspectors are invited to modify their roles to test the lived experience of the teacher and to try to help them find the solution themselves, on the condition that this goes in the direction of the reform. Controlling subjectivities is not far away.

Why support? The last formula, that of supporting reform, cannot help but recall support for the dying, where the latter cannot choose between life and death but can only die with dignity, without suffering and in peace. If we make a bold comparison with the innovator, this means putting oil on the gears for enacting the reform without obstacles or conflicts. Promoting innovations returns to inviting innovators to enter into a mold compatible with the axes of reform, at the price of an adjustment carried out through support[53]. Every innovator who takes a differing or opposing path will see themselves not supported and their innovation not promoted.

Support, sometimes accompanied by allowances in timetable or in ways of obtaining material or financial aid, allows the steering of an innovation towards the enactment of the reform. Support thus becomes the strong arm of educational policy, and loses its aspect of helping, of non-intervention in the choices of the supportee.

53 It is not far from support to manipulation, to the extent that support is inevitably an intervention in someone else's activity, leading the latter to do things that they would not have done without this support.

However, support stripped of proactive intention allows control by proximity; it facilitates transformation in relying on current practices and new experiments. Based on volunteering, personalization and proximity, support can be a largely facilitating mode of change in professional representations and practices. The competences of the supporter are at work here.

But this does not mean that it must be rejected because, as we stressed at the beginning of the work, innovation may be born from reforms – the two are not incompatible! In fact, organizational or structural reform can engender innovation; the reverse is less frequent.

Vial and Caparros-Mencacci[54] have identified seven essential ideas concerning support after a study on people responsible for support in the educational world:

– it is a particular form of helping practices;

– it is a professional activity which escapes purely cognitivist conception (there is a relational component);

– it is a practice of evaluation;

– three fundamental processes are at work: orientation through action, benchmarking and problematization;

– it is a specific kind of relationship;

– it gives directions for action;

– it is a form of professionalization of the supporter.

One last question to raise: can there be such a thing as a professional in support, and does it not rather entail improvisation, bricolage and willingness from the supporter? What are the competences needed to support innovators, and is there such a thing as training, or do we think that this action requires designated or self-designating people as supporters according to the conceptions of this activity by the same people who make the designation?

54 VIAL M., CAPARROS-MENCACCI N., *L'accompagnement professionnel ? Méthode à l'usage des praticiens exerçant une fonction éducative*, De Boeck, Bruxelles, 2007.

There has been no evaluation up until now of the actions and effects of support for innovators.

In France, at the academic level, there are academic advisors to help the development of innovation and experiments (CARDIE). This help can take the form of support in a continuous process, putting innovators into a relationship with other actors such as trainers.

At regional level, selected innovators can benefit from support. The latter is conditional on a series of steps leading to this official recognition.

Figure 2.3 illustrates this process.

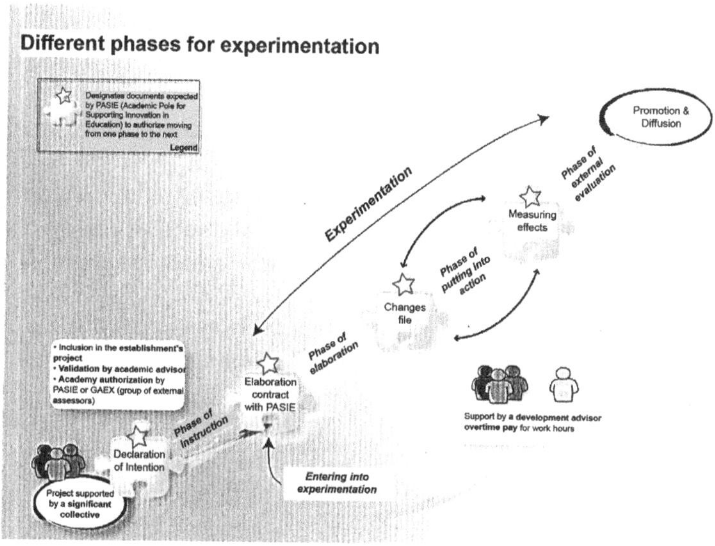


**Figure 2.3.** *Different institutional phases with which to experiment*

In the schema provided by the Academy of Lyon, we see the (obstacle?) course that an innovating team must follow in order to become recognized and to benefit from recognition, possibly including allowances and support from trainers and researchers.

This schema should be read from bottom-left to top-right. First, the innovation must be transformed into a project submitted to the Board of Directors of the school and thus be registered in the school's project. Then, the innovation undergoes transfer to the academy level, which examines the written project (requiring the filling out of a standardized file) and then a contract is drawn up for the beginning of experimentation, ending after 2 years with a double, internal and external evaluation (the latter often carried out by inspectors).

Thus, what is left of the novelty of the innovation at the end of this course? This support is often obtained at a steep price, and it is in this sense that few innovations have thus been collected and followed up in relation to the desire for innovation from teachers, while paradoxically we still maintain a discourse complaining about the immobility of teachers!

## 2.8. The place of pedagogic movements and great pedagogues

Most innovations do not appear out of thin air; they are influenced consciously or unconsciously by conceptions of the process of children's learning and ways of helping and developing their educational potential.

Without returning to the pre-Revolutionary era evoked by Ariès[55], where the place and image of children changed greatly within the family and society, we can observe a revolution in the place and role attributed to them.

Children were for a long time considered to be small adults (see certain paintings of the 18th Century), acting as a source of labor for the family. They were seen as an inferior subject, owing obedience and exercising invisibility, only to become in our modern Western societies the center of our preoccupations, and the center of our schools. In other words, children moved from the status of individuals of little importance to that of bearers of the future of humanity, requiring careful attention. Children became the object of investment in the future and must thus be raised in joy, pleasure and benevolence.

55 ARIÈS P., *L'enfant et la vie familiale sous l'Ancien Régime,* Points, Le Seuil, Paris, 1960.

Current innovation is strongly inspired by this image of the child in society, where it is appropriate to find the best ways to "naturally" promote the development of young people, in all their potential. This is compounded by the deep wounds in Western societies from the two great World Wars, which have driven them to flee misery and suffering for a world of brotherhood and humanity, in other words a world of respect for others, diversity, autonomy and democracy.

Moreover, it was in France (between 1917 and 1939) that a preoccupation developed with the methods of teaching young people. This preoccupation was not devoid of ideological currents, as all education carries within it an embryonic vision of the desirable society.

Without dwelling on this array of people leaning over the cradle of future citizens, we can distinguish three categories of them: one being the instigators (those who have suggested what must be done for the youth), another the great pedagogues (those distinguished by a theory of the methods of teaching, often with one or more aspects drawn from the instigators) and, finally, the pedagogic movements structured according to common concepts, in struggle against identified enemies and intervening in schools, outside schools and around schools.

### 2.8.1. *The instigators*

These are generally philosophers, researchers or religious figures, such as Ignatius Loyola or Jean-Baptiste de Lasalle, Rabelais (the Abbey of Thelema), Rousseau, Comenius, Montaigne, Piaget, Dewey, Confucius, Wallon or Skinner. They advocated strong ideas on the process of intellectual, social or personal development of young people and on ways of aiding their development, while socially integrating them.

How can we not see Rousseau's ideas emerging from certain current innovations? How can we not see the influence of Piaget's structuralism in other innovations, or even a little of Skinner's behaviorism in the use of goals, which was an innovation of the 1960s?

These instigators do not share an identical approach, but they all have the merit of pointing out the complexity of the development of young people and stressing that ways of addressing it may be different when seen through different theoretical lenses. Each provides an intellectual consistency to understanding the child, which is difficult to deny, but whose limits must still be pointed out.

### 2.8.2. *The great pedagogues*

These are in general women and men who have enacted in a concrete and systematic manner, in schools, more or less the thoughts of the instigators, while using their own filter and adding their own ideology, that is, their social vision and procedures likely to comply with it. For example, Claparède, Pestalozzi, Ferrière, Montessori, Cousinet, Freinet and Oury in institutional pedagogy.

They have set out pedagogical methods and have enjoyed more or less faithful successors who have promoted their principles of implementation. They have sometimes been forced to leave the public school system due to public incomprehension and political positions concerning schools, which still envisaged them in the manner of a barracks or a religious convent. Many books have been published on their work as well as training schemes set up for willing and interested teachers.

It is useless to remember that these "great" pedagogues dealt with current pedagogies considered to be innovations, which may appear to some to be recent. The example of Cécile Alvarez[56] and the natural laws largely inspired by Maria Montessori certainly fall within this category.

### 2.8.3. *Pedagogic movements*

Finally, there are the pedagogic movements. They are structured around a concept of education, based more or less on the two preceding elements: one or more instigators and one or more great pedagogues[57].

56 Cécile Alvarez is a young teacher who, with the support of the Chatel ministry, tried to innovate in a public school. She could not continue her experiment and left the public sector for the private.

57 We are not taking into consideration the different associations of teachers, often structured along disciplinary lines like, for example, the Association of Teachers of Living Languages or the Association of Literature Teachers, whose goals are different as they are organized around the didactics of one discipline or a set of related disciplines.

There is a great diversity of these movements in France, all in the form of acronyms that we will not expand upon (such as AFL, CEMEA, CRAP, FOEVEN, Francas, GFEN, ICEM, FOEVEN, USEP, OCCE and EPI). In 1985, the National Institute for Pedagogic Research (INRP) tried to group them under the title of CLIMOPE (Liaison Cell for Pedagogic Movements) with a directory of these organizations. The experience does not seem to have lasted, as ideological differences exceeded the possibilities of working together. As we noted in the Introduction, once more we have evidence that school is criss-crossed with ideologies and that, under apparently common ground (good teaching of students and pedagogic approaches, which are said to be close, such as student activity, its universality and cooperation), there is clear divergence.

These movements are incubators of innovations; they support training and stimulating activities through their very active networks. Some possess research centers and are linked to research laboratories for social sciences and humanities.

### 2.8.4. *Pedagogic currents*

Pedagogic currents play a central role in innovation, not like the preceding organizations with their active participation in stimulating change in teacher practices, but because they have intersected with them. In fact, a pedagogic current is an explanatory theory of forms of student learning and therefore of their associated pedagogic situations.

Houssaye[58] was one of the first researchers to try to clarify what professional action takes place in a school to allow students to learn (Figure 2.4). This is what he calls the "pedagogic triangle", along which it is possible to place the various activities performed by teachers: on one side of the isosceles triangle, the link between knowledge and teachers is seen as a concern for teaching, putting the emphasis on knowledge and its introduction by the teacher (forgetting somewhat the complexity of the profile of students); on the other side of the triangle is the link between knowledge and students, stressing the forms of student learning whilst teachers are less present; finally, on the third side, the link between

58 HOUSSAYE J., *Le triangle pédagogique,* Peter Lang, Berne, 2000.

students and teachers involves the social and affective relationships, which link students and teachers for the sake of education.

If this triangle may seem simplistic and as if it does not correspond to the complex, changing reality of all the teacher's pedagogic practices, it has the merit of pointing out the dominant issues in certain practices and working to fill certain gaps. Houssaye clarifies that often these connections happen in pairs, rarely all three, one of the poles playing the role of the "dummy".

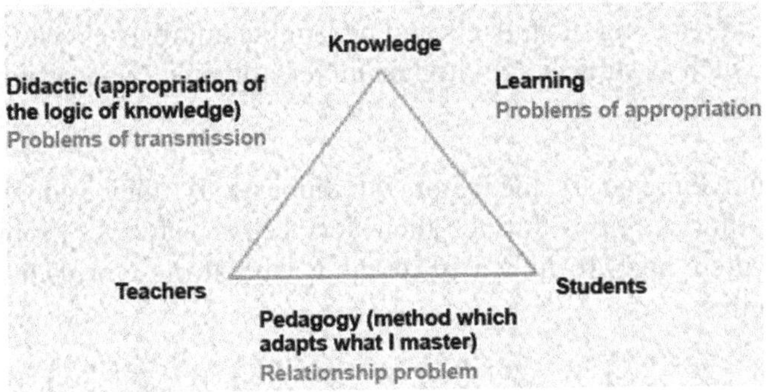


**Figure 2.4.** *The pedagogic triangle according to Jean Houssaye*

Starting from this triangle, it is easy to locate some of the theories proposed about learning, theories which have greatly inspired some innovators, in a successive stream of new theories. One theory is chased by another. It arises from the work of thinkers and researchers who interpret, based on their data and the progress of research results, a form of the development of learning.

Table 2.4[59] summarizes the four major theories that have intersected with the inspiration of innovators:

– the constructivist approach based on the idea that the individual constructs the knowledge they acquire, and that this knowledge is not simply swallowed as-is;

59 Table taken from the work of CARRE P., CASPARD P., *Traité des sciences et technique de la formation*, Dunod, Paris, 2011.

– the cognitivist approach based on a principle of structuring intelligence through problem-solving;

– the humanist approach, more based on allowing free rein to an individual who, under light guidance and flexible and benevolent support, will be able to develop all of their learning potentials;

– we might add an approach which is more linked to the social domain, the sociocognitive approach, which indicates that learning is facilitated in peer exchanges under specific conditions of conflict.

| Constructivist | Cognitivist | Humanist | Sociocognitive |
| --- | --- | --- | --- |
| Dewey, Piaget, Vygotsky | Bruner, Chomsky, Gagne | Rogers, Maslow, Deci and Ryan | Bandura, Zimmerman, Dweck |
| Construction of representations based on experience | Process of treatment of information | Personal action of self-realization | Modeling, regulation, self-management |
| Internal development | Cognitive structuring | Affective and cognitive needs | Self-/social interactions |
| Allowing access to meanings by the subject | Structuring the content of the learning task | Supporting personal development | Facilitating self-regulation of learning |
| Creating situations fostering the progressive construction of structures for comprehending reality through problem-solving | Aiming at the improvement of cognitive capacities and strategies through training and structuring of mental operations at a given level | Privileges personal development, flourishing and self-actualization and autonomy with regard to the relationship of assistance | Facilitating the adoption of self-regulated learning behavior, thanks to direct or vicarious experience and the development of a feeling of personal efficacy |

**Table 2.4.** *The four major pedagogic currents*

These approaches are sources of inspiration for innovative teachers who find, through these theories, matter for clarifying their pedagogic practices and, above all, for formalizing them by putting them in the field of the scientific and the explicable, indispensable elements for promotion of their innovations.

## 2.9. Training innovation in schools

### 2.9.1. *There is innovation, and innovation*

As we have seen in the preceding chapters, there is innovation, and then there is innovation. If it means putting a project pedagogy in place, we can estimate that it has been more than 40 years since this was introduced to the French educational system and that there is no real danger. We would say that such an innovation is rather in a phase of routinization. If we had to draw up a map of declared or undeclared innovations, we would doubtless find that the number of innovations involving project pedagogy would be very high!

But, if it means recklessly putting in place an innovation, such as multi-age classrooms, even if we know that rural classes have had to manage this kind of environment, it is not necessarily obvious how to manage a multi-age class of 35 students and, in this case, imagination must be strongly involved in organizing mobile groups for mutual aid in learning, using self-evaluable goals.

In other words, all innovations are not the same, and yet many arise at the same time in a sort of implicit necessity. These innovations are more or less transgressive, even if their institutional framing is vigilant. Establishing a project pedagogy in a discipline at a single level of students is less disruptive than allowing students to leave the institution to create files on themes, which are relatively wide and involve several disciplines. And we are not talking about innovations which, beyond pedagogy, lead to organizational and structural changes.

Innovation may be situated at another level: that of training trainers in innovation, so that they may in turn train teachers in innovation. The initial hypothesis rests on a homothety or isomorphy of training (experiencing what we want teachers to experience so as to train them) and it is here that the question arises of how to train teachers in innovation.

How to carry out this training, because even if we know how to innovate, have experienced it and retrospectively evaluated it, are we capable of training another to innovate? A good butcher might not be able to train another as a butcher. It would be doubtless necessary to didacticize the knowledge gained about innovation. For this, it would be necessary to identify and work on this knowledge, to be able to develop it in the future teacher. In other terms, it requires didacticizing the competences of innovation[60].

In teacher training activities, trainers have experimented with their forms of innovation. They hypothesized that, as they have all experienced innovation in training, they know in turn how to enact innovations, which train students to innovate (but for newcomers, what is it new in relation to? Their past life as a student? In short, it may be an innovation for the trainer, but not for the trainee, even with students who are not conscious of innovating).

Giglio and Perret-Clermont[61] put in place a method of innovation in an elementary school class with teachers in a music course. It consisted of students in groups creating a musical piece using a collection of musical instruments.

This creation had three moments: a moment when the group begins work with its hesitations, momentum, refusals and rivalries. A moment when the teacher moves between each group and questions them to discover difficulties and provide ways of surmounting them. At this time, the teacher offers some knowledge, which may unblock the process of creation, due to the students' lack of expertise. Finally, a moment of coming back to the whole class with a double exposition: first producing the piece of music before the others and discussing it with commentary from the teacher (adding their experience) and then explaining how the group work was carried out.

The two authors, a researcher and a trainer, subsequently questioned the teachers who created this innovation and whose object was to develop creativity among students. They unanimously agreed that they tended to interfere too much, to anticipate students' questions and to want to provide answers which did not always correspond to the students' issues. We might here mention what Rousseau wrote in *Émile*: "Young teacher, I am setting before you a difficult task, that of doing everything without doing anything

60 We dealt with this question in the work CROS F. (ed.), *L'Agir innovationnel: créativité et formation*, De Boeck, Bruxelles, 2007.
61 GIGLIO M., PERRET-CLERMONT N., "Créer un objet nouveau en classe. Un dispositif d'innovation pédagogique et d'observation", *Éducation nouvelle*, pp. 211–237, 2016.

at all", thus stressing that the most important thing in training a person is their engagement.

The pedagogic method relies on arousing the creativity of students, while in no way abandoning the course's content and goals. This example is very instructive for the training of teacher trainers, because if the goals are to develop the creativity of the future teachers, offering them an unprecedented situation where they have to get groups of students to use their collective imagination to produce an object which is unknown at the start (a piece of music with all the forms of musical instrument available) is original.

We might imagine means of training future teachers inspired by this innovation in an elementary school. For example, instead of suggesting a portfolio on the Internet to trainees, one which undoubtedly displays their competences, it might be possible to ask trainees to imagine a medium which would be able to display their acquired competences to someone else. Trainees would see that they had a tendency to be more interventionist than necessary, and would respond only to the extent of questions raised by the groups. Basically, we are not far from good old non-directivity, on condition that it is framed by a controlled didactic process. The learner becomes the author of their learning in an activity requiring creativity and the construction/integration of new knowledge.

Not every innovation has this bonus value; under its cloak of novelty it may conceal habits which escape the teacher and invalidate for the student the possibility of being truly creative.

### 2.9.2. *Well-identified competences for the innovative spirit?*

What, then, are the required competences for an innovative teacher, compared to a teacher whom we would consider non-innovative? What are the qualities of an innovator? Do we even have to abandon academic training?

Training for what? A definitional consensus on innovation must be established at the outset. Once the object of training has been identified, how to proceed to train and evaluate the required competences?

Is innovation a different form of practice? In what way? Against routine, repetition, tradition, the past, habit, repetition, the academic, the masterly?

From 1 July 2013, in France, a decree has specified the competences necessary to perform the profession of teacher. Whilst the previous syllabus mentioned competence in innovation, it has disappeared from the most recent version. Why? We can say that the faculty of innovation is no longer a competence pertaining to the profession, but that it covers all of them. An interview in the same year with the manager of the Directorate-General of Teaching in Schools confirms this impression, as she explained that "tomorrow's teacher must develop their curiosity and their creativity, be in a position for research and experimentation and must innovate on a daily basis".

In other words, a teacher is essentially a person who innovates in their professional practice. This is no longer breaking news but, at the same time, obvious and a banality. The profile of a new teacher is changing: they innovate constantly, sometimes without knowing it.

It is thus within all the content of training itself that innovation develops: it is therefore enough to allow the teacher to problematize the professional competences, to continually imagine new ways of learning for their students in collaboration with other teachers and staff of the institution, and to involve parents in it, to be an innovator or, more exactly, a high-performing teacher.

However, we intend to identify the competences intersecting these contents, which are prone to developing this spirit of innovation. As Tedesco (1994) puts it, "it is more important to generalize the capacity to innovate than to generalize the innovations themselves".

### 2.9.3. *Intersecting competences for developing innovation*

Preliminary considerations:

– Teacher trainers currently have in mind that the content of professional competences must be imbued with the spirit of innovation, that is to say, requiring imagination, creativity, the unusual and even sometimes transgression (out-of-the-box thinking)[62].

– Even if such a statement is agreed by the majority of trainers, we must nevertheless remember that French national education has obligations in terms of professional ethics. It is thus appropriate to innovate in the framework of the prescribed exercise of the profession, in a recognized institution, with other people.

– There is innovation that the teacher can introduce alone in the context of their class, and innovation pursued by a collective, which the institution strongly encourages. The profession of teaching is thus not a solitary profession, but a collective practice that is ceaselessly renewed.

We gather under three major headings the intersecting competences, which may develop the capacity to innovate, as discussed by Tedesco[63].

#### 2.9.3.1. *Cognitive competences*

– *Creativity* (out-of-the-box thinking, new ideas). Psychological studies show that people whose thinking is divergent (that is, involving new interpretations and deviating from customary uses of objects or social relations) are inclined to be innovators. This competence is the basis of innovation: without creativity, no innovation. This does not mean creating new objects at any price, but allowing a recomposition of customary uses of things, activities and social relations.

– *Imagination* (thinking the impossible, imaginary). Daring to think outside usual frames of thought, being freed from temporarily binding frameworks and being able to communicate it.

– *Original reorganization of existing things*. Derailing habitual thoughts, playing on words, on situations.

62 WATZLAWICK P., WEAKLAND J., FISH R., *Changements, paradoxes et psychothérapie*, Points, Le Seuil, Paris, 2014.

63 Competences are a mobilization of a set of capacities in a given professional situation with its affects, emotions and social dimension. Training for a competence requires developing capacities, but the latter cannot be competences in themselves because they must be mobilized during real situations.

#### 2.9.3.2. *Communicative competences*

– *Formalization of practices.* An innovation which is not communicated is an innovation which only exists for the innovator. Now, most frequently, the innovator wishes to make their discovery known, to share their success. This is a central point of innovation if it is to have a life beyond its author.

– *Argumentation, persuasion and conviction.* Innovation which is communicated must be defended, even to the detriment of the privilege of authorship (accepting the erasure of its origins so that the innovation may continue to live).

– *Multiplying channels of communication.* Formalizing the innovation using a multiplicity of media for different publics, so that the innovation not only becomes known, but adopted (see the effect of "translation").

#### 2.9.3.3. *Personality competences*

– *Tolerance for frustration* (compromise). Accepting that the innovation will not always be well received and free from challenges or criticism and that it would be appropriate to create compromises resulting in the innovation developing in a more attenuated or slightly modified form.

– *Letting go.* That is, remembering that it is the student who learns and doing everything to allow their full engagement by not intervening too much, just as much as necessary.

– *Enthusiasm and dynamism* (risk and doubt). The innovator's motivation must be strong so that they are persuaded that their innovation offers the best solution to the expressed need. The danger here is that of militancy. They must therefore keep a cool head and remember that their innovation may have limitations.

### ***2.9.4. Ways of training trainers to develop a global capacity for innovation***

We have seen above how a pedagogic situation may lead to developing creativity in the trainee. The situation in the music course is one example among many others.

One of our students[64] conducted research on the training of future teachers. The goals of this course involved teaching writing to the student. Rather than giving a course on the subject, the trainer placed large sheets on the wall and handed out large brushes and paint. The instruction was for every trainee, one after another, to write their signature. Everyone watched each other's production and drew their signature in a broad motion, standing upright, in front of the others standing around them.

This new experience, aside from revealing the global physical aspect of handwriting (breath, grasp, precision of gesture) produced original signatures in very different colors with thicker or thinner brushes. The trainer had never considered this pedagogic situation as developing creativity among future teachers and yet the fact of commencing an unknown, unprogrammed, original activity led each one to imagine and position themselves in a four-dimensional space in a sort of dance of the brush.

What do we want to say with this example and the preceding one of the music course? Simply that the trainer creates situations where they themselves do not control the outcome and where the freedom of the trainee, although well framed in precise and clear instructions, is open to a multitude of possibilities, themselves enriched by the collective. The activity of the trainer, like that of the trainee, is to distance themselves from interventionism, from the idea of identical paths or of a best path, the others being only diversions.

64 MARTIN S., *Approcher le re-vécu d'une expérience graphique située en formation des enseignants*, Conservatoire national des arts et métiers, Paris, 2015.

# 3

# Innovation in Schools: Issues for Policy and Governance

## 3.1. Innovating schools: learning schools?

Innovation in schools is often seen as the creation, either of a teacher or of a teaching team within an establishment, of which certain members are not necessarily involved in the innovation. Pessimists believe that such innovation has little chance of having a strong and sustainable impact on the whole school establishment. They strongly encourage all staff in the institution to innovate, that is, to get involved in constructing new practices and new ways of learning for students, including parents. In other words, accepting these two concepts (an innovating school and a learning school) implies that the whole of the school must be involved, in a systematic way.

The distinction between innovating schools and learning schools is not always very clear, and depends on the language game being played by the speakers. Imagine a school where all staff innovate, that is, they create and invent ways to do things differently with the goal of improving student learning. Would that not be at the same time both innovating and learning?

The notion of innovating schools, however, predates the use of the notion of learning schools. It seems that an innovating school was seen as a nursery for many new actions, more or less coordinated. What introduces the notion of learning schools is the fact that the role of conductor of this orchestra of innovations is more emphasized as an organizer of innovations, the force behind their promotion and evaluation. This notion of "learning" plays down the concept of individualism even more, which was positively promoted until recently, in favor of working in teams and as a collective (with the idea that the whole is greater than the sum of its parts).

Moreover, works written about learning institutions put their emphasis on the actual management of the organization[1], on its *leadership*, on the coordination of a collective intelligence and its exploitation through strong interaction with the continuing training of staff. Capitalization on innovations is intense here, as a result of these new ideas transformed into innovative activities.

As much as innovating schools are centered on new modalities of pedagogic practices, they are equally at the service of a process which goes beyond them, which is concerned with results in terms of change in the institution itself, in its structures as well as in the activities which are carried out there. The cognitive and regulative aspect is foregrounded here.

We can sum up the difference as in Figure 3.1.

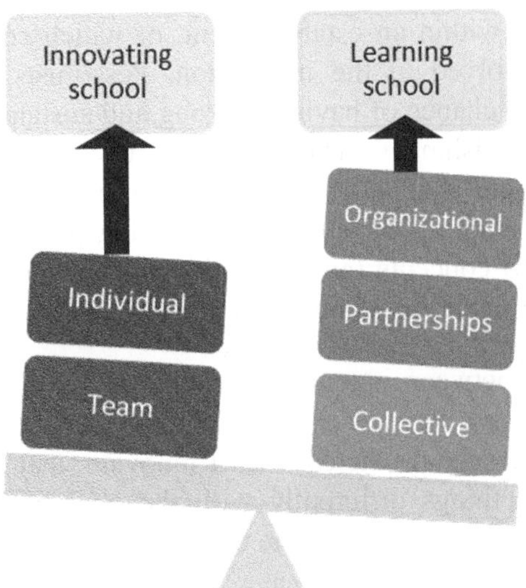


**Figure 3.1.** *Differences between an innovating school and a learning school*

1 BOUVIER A., *Vers des établissements scolaires apprenants. Perspectives pour la conduite du changement*, Editions Canopé, Paris, 2014.

The balance between the two dips in favor of learning schools, if the main goal is to proceed to macroscopic change in the educational system. Furthermore, it is interesting to note that a primary author of works on this topic, Alain Bouvier, a former Rector of the Academy, discusses the managerial transformation of the educational system in his most recent book. Moreover, he recently published another work, more of a manifesto concerning the management of reforms[2].

On this topic, he formulates 10 principles for reform in schools and developing so-called learning schools:

1) develop a systemic project, with a vision and goals, based on the situation on the ground;

2) individualize students' learning and their journey;

3) increase the number of days in class;

4) in professional teaching, reverse the current links to employers and increase the employability of students;

5) intensify the professional development of actors through innovation;

6) allow parents to become more involved;

7) set up new forms of territorial governance;

8) establish self-evaluations of schools which will be made public;

9) give confidence to actors;

10) reduce central administration to 20% of its current workforce, while changing its role.

Such a macroscopic vision is a good illustration of concern for the direction of the educational system, while still leaving spaces of freedom for the schools themselves.

At the international level, these two concepts are often discussed, particularly at OECD symposia which bring together very diverse countries: what do we mean when we talk about a learning school? A particular activity carried out by identified actors in a school? An overall evolution of the institution as a whole? Performance results which pragmatically engage with civil society as it exists?

2 BOUVIER A., *Pour le management pédagogique*, Berger-Levrault, Paris, 2017.

Continuing training cannot avoid reflecting on these different concepts and on the "magical" notion of a learning school; that is, one where all staff are involved in discovering themselves through creativity which is stimulated, for example, by the parameters we mentioned at the end of section 1.1.4 (Figure 1.2).

## 3.2. Evaluating innovation in schools

Innovation, due to its risk-taking and uncertainties, is often questioned about its effectiveness. Questions pour in such as: "Great, you've had fun with the area of student learning, but what are your results?"

This is a way of minimizing the presence of pleasure, joy and play in learning, still seeing it as involving austerity and cognitive activity stripped of positive affect[3]. It is also a way to oppose any possible trouble caused by uncontrollable innovation. However, an innovator cannot just do what they like and it is essential that they be accountable.

In addition, many evaluations of innovations took place in the 1990s, carried out by the research service at IREDU[4] in Dijon. The results of these evaluations were in general mixed, not to say negative. This might lead us to judge innovation as, if not ineffective, in any case not dangerous!

Two explanations are possible: on the one hand, motivation, momentum and the dream of a better future are difficult to assess; on the other hand, evaluation is carried out in the same way as evaluation of a common pedagogic action, that is, the extent to which students have learned, based on classical criteria of acquisition.

Such an attitude makes us think of La Fontaine's fable, where the stork and the fox, each inviting the other to dine at their home, are both amazed that the other guest cannot eat anything! The evaluation of innovations is more complicated and most of the time, leaders have passed this on to the

3 On this topic see the work of DEHAENE S., *Apprendre à lire. Des sciences cognitives à la salle de classe*, Odile Jacob, Paris, 2011.

4 Institute for Research in Education at Bourgogne; sociology and economics of education.

innovators themselves, asking them about the effects and impacts of their innovation, which most of the time ends up in qualitative self-appraisal[5].

The hot topic is whether or not to redirect innovation to the goals of student learning as defined in official texts, whereas some innovations are seeking a nourishment of spirit which is seldom sought after in schools (creativity, freedom, cooperation, interdisciplinarity, collective creation, participation of parents, etc.).

As we stressed in our definition of innovation in schools, it is a social innovation, that is, an activity which responds to a need not fulfilled by the institution. So how can the institution evaluate activity towards a goal which it does not develop? Innovation is often a response to a deficiency in the institution on two levels: on the one hand, the process of acquisition of learning and, on the other hand, the behavioral conduct of students and teachers.

If the innovations in question clearly target the goal of improving learning in this way, thus lining up with the official state of knowledge, such assessment is quite possible. But if the innovation is more challenging and pursues goals which are not valued by present-day schools, the usual form of evaluation will lead to disappointing results. This is doubtless what happened with the evaluations carried out by IREDU.

This is also what often happens, since innovation is carried forward by a strong will, a desire to do something that will make things better. But this "better" is full of subjective projections of such a kind that the objectives are not totally explicable. Now, leaders request that the innovator formalize in advance for them the goal or goals which they pursue. If they want help, support, institutional recognition or underwriting, they must transform their innovation into a project with clear, scheduled goals and adequate means of evaluation.

A simple example may demonstrate this tension: a team of teachers decides that the final-year students at a junior high have some problems with reading. They thus decide to put in place an innovation consisting of inviting students to read short stories or news stories and, the next week, to tell them to others.

---

5 It is enough to read the presentation sheets for innovations on Expérithèque to see evaluations carried out by the actors themselves, who are not in a position to go on to a critical analysis of their own innovation. Evaluations thus result in nothing more than self-appraisal or reports which are largely subjective and not very representative.

The students enjoy this way of doing things, get on board with the project and increase the volume of their reading each time, readings that they get from the documentation center. After 6 months, the teachers proceed to an evaluation of the quality of reading through a standardized test which seems appropriate. Imagine their surprise to discover that the students have not progressed in reading as they hoped but, on the contrary, they have greatly progressed in oral expression, in the presentation of texts! The central goal achieved has become that of oral expression and exchanges between students in their conversations and narrations. In other words, the originally stated goals are not those achieved by the innovation and in fact the opposite situation needs to happen, that is, defining *a posteriori* how these activities have led the students to such and such a new realization, producing a new goal.

This evaluation of innovation is inscribed onto a complex base of evaluation, because what does evaluation mean in education? Régnier[6] wrote that "evaluation in education means bringing a distanced and critical view, making an effort at systematic lucidity whose purpose is to provide information which helps us make decisions. The quality of the evaluation rests above all on the degree of adaptation between the objects of evaluation and the evaluative situations". Here, the object of evaluation is innovation as a whole, and may involve many different sub-objects: value added, novelty, change in the role of teachers, change in student behaviors, etc.

There are in fact several ways of evaluating an innovation, but the most-often requested is that of effectiveness with regard to the institutional goals of acquisition of competences by students. However, we must not neglect other types of evaluation which help in the development and maturation of an innovation (process evaluation, diagnostic, formative, summative, self-evaluation, co-evaluation, etc.).

An analysis grid for candidate files was developed by the French Ministry of Education for choosing innovations in 2016. To help with the qualitative analysis of fact sheets concerning so-called innovative actions, the DRDIE provided CARDIE with a reading grid.

6 REGNIER C., *L'évaluation dans le système éducatif*, CNIRS, Paris, 2002.

Six domains of innovation were identified covering the whole area of innovation, according to qualitative degrees ranging from an innovation at its beginning to a more inclusive innovation also involving teacher training, summed up in Table 3.2.

Such evaluation displays an intention of steering innovation through evaluation. Only those innovations are retained which correspond to the requested profile. The absence of a place for research in this type of evaluation leaves the door open to authoritarian diversions, while ignoring the positivist magic of the *evidence-based policy* process!

| | A | B | C | D | E | Dynamic |
|---|---|---|---|---|---|---|
| **Domain** | *Positive reaction by students* | *Significant improvement in student achievement* | *Professional development of teachers* | *Change in the organization of work* | *Collective competence, sharing roles and responsibilities* | *Level of development* |
| **1st degree** | The action is not very relevant to professional ethics | Not evaluated for now | Little writing or communication on the action itself | The action is in the class framework | The action involves a teacher in their class | Emerging practice |
| **2nd degree** | The action potentially concerns every student | Progress has been noticed in the working environment | The action is visible: the team communicates an analysis of its activity | Some *ad hoc* changes to schedules | A team is formed for a sustainable action | Innovative method |
| **3rd degree** | The action particularly deals with students in difficulty | Progress is visible for most students | A means of analysis is set up: co-observation, external analysis or support group (CARDIE, IFE) | Use of time has changed to allow an action of greater scope: groups of students vary | A project coordinator or leader exists. A regulatory body in place | Learning organization |

| | | | | | | |
|---|---|---|---|---|---|---|
| **4th degree** | The action integrates formal (scholastic) and informal dimensions; it takes into account, for example, the feeling of well-being | All students have progressed | Staff become resources for internal co-training; a regulatory group is in place; the team is a resource for other teams | Variable-time modules in non-school format appear; the team is larger | Action is coordinated with management; it is included in the school project | Innovating school |
| **5th degree** | The action explicitly includes an overall impact on students (belonging, well-being) and creates its evaluation with them | All students have progressed. Visible results, in exams as well as insertion | Staff participate in professional networks; the school becomes a trainer and joins the ESPE (Higher School of Teaching and Education) network | Action requires specific competences and sometimes specified jobs, depending on the organization of work | The action is at the heart of the institution; its dynamic involves a large part of the team, as well as partners | A training establishment or an experimental structure |

**Table 3.1.** *Example of an official evaluation grid for innovation in schools*

Innovation is, by definition, the introduction of something unprecedented, of value added with respect to what was envisaged. The particularity of the assessment of innovation might center on this novelty. Taking the unprecedented into account leads to innovation in the assessment of innovation! If we return to Alain Michel's definition, "innovation is an invention which has succeeded", but in what and for what?

## 3.3. Innovation in schools and research

Current tendencies in supporting innovation in schools strictly relate it to research on education in all of the disciplines that pertain to it, particularly in neurosciences which currently have the wind in their sails, and in social sciences and humanities. There are three ways of linking innovation and research:

– the first is to consider that all change driven by policy (*top-down*) can only be enacted and accepted through the famous *evidence-based policy*, that is, through incontestable and uncontested research results. Now, it may happen that several studies on the same subject will not lead to the same results, simply because their research tools, collection of data and interpretations are different. Research in social sciences, for example, faced with a subject as complex as school, cannot function in terms of evidence and truth. However, it is useful to demonstrate the complexity of the subject and avoid possible deadlocks;

– the second involves researchers themselves carrying out research in close cooperation with teachers and tends to demonstrate the effects of changes in teacher practices on the performance of students (qualitative and quantitative). The direct relationship set up between researchers and teachers allows mutual clarification of the situation of innovation and advances in direct understanding;

– the third is to train teachers in methods of research without yet becoming professional researchers, but in order to possess the methods and attitude of research which allow them to question their change of practice in a systematic and rigorous manner, sometimes aided by theoretical readings (which trainers may recommend to them).

### 3.3.1. *Evidence-based policy*

This approach takes innovators as a field of research with, to a greater or lesser extent, their participation. The underlying idea is that providing evidence which is incontestable and communicable in all environments will allow support for good innovations, those which make progress towards proven development in student learning. Now, research is neither immune to the ideology running through the school, nor to the values which underlie it. Providing evidence is not enough to grasp the specificity of innovation in schools. The question raised is that of moving from the results of research based on evidence (thus linked to causality) to change in professional practices. The advantage of such an attitude is in trying to depoliticize and objectivize educational and pedagogic choices.

### 3.3.2. *Action research or training research*

This involves research where the protagonists are researchers and innovators who work in the same direction of clarification, with the assumption that researchers understand the world of research, but are far from understanding the world of teaching. Moreover, innovators have things to say on their activities beyond simple observation. The meaning of the action cannot be dissociated from that of the actor who conducts it. This action research or training or development research still has its limits in the generalization of its results[7]. It often results in the production of resources which are then made available to teachers.

The creation of Associated Sites of Education (LéA) by the Lyon Institute of Training and Education (IFE) is part of the continuation of what was done at INRP, that is, giving the opportunity for researchers, actors and leaders to think about and to test possible concrete responses. This requires collaborative work between researchers and actors on a research project which most often produces exportable resources. This link between the two worlds (research and the teaching profession) inevitably results in giving research a utilitarian flavor, more immediate than would be the case for long-term research producing decontextualized theoretical knowledge[8].

The creation of the Carnot Institutes of Education also relates to this research approach, which resembles poles of competitiveness for education based around a well-defined subject.

### 3.3.3. *Research as a competence of innovators*

Besides all that, the spirit of research, much like the process of learning which is advocated, for example, by the innovations enacted by the international association "La Main à la Pâte" (Shoulder to the Wheel), is to consider the professional development of every teacher according to a

7 CROS F., "La recherche en éducation: une nouvelle forme d'accompagnement de l'innovation, entre acteurs et décideurs politiques", *Accompagner les réformes et les innovations en éducation,* L'Harmattan, Paris, pp. 101–124, 2004.

8 Many studies have been carried out on different types of research, particularly on the research called "professional". On this subject, read the publication of AFFUTS (French Association for the Development of Research in Social Work): *Quels modèles de recherche scientifique en travail social?* Presses de l'EHESP, Paris, 2013.

research method consisting of experimenting, making hypotheses, examining the effects and proceeding to reinforced adjustments.

What we might call researcher competences for a teacher correspond to a conception of the teaching profession inherent in a process of continuous learning/training. The methods of this approach are those of the research process which are in fact identical to the approach we use with students, that is, proceeding through the scientific rigor of hypotheses, attempting to verify certain facts and interpreting them in order to then be able to integrate them into the development of professional competences. Training throughout the whole professional career of the teacher is thus implied in this perspective of a research process.

Since the creation of the IUFM (University Institute for Teacher Training) in 1991, training has relied on research to the point that future teachers are asked to write a research report about professional questions that they have concretely encountered during their sole-charge internship.

But there is research, and then there is research. It is not a matter of turning teachers into researchers, but to developing in them the research competences which continually enable adjustments and creative professional activities, maybe even to use the publications of education researchers.

The most common idea of a researcher is that of a research professional, trained for this role which becomes their career. They are often professionally attached to a laboratory and must follow international conventions of publication.

The second conception of research carried out by teachers themselves recovers the idea of an intellectual approach which responds to the imagined procedure of POHEC – Problem, Observations, Hypothesis, Experimentation (this is where innovation comes in!), Conclusions and results. It involves an attitude of the same kind as the new pedagogic approaches to science and technology in schools such as "La Main à la Pâte" as mentioned previously, or that of "adventure scientists", consisting of no longer swallowing the theories or laws provided by scholars, but trying to begin with hypotheses to verify, constructing a simple experimental method or one inspired by experiment, and drawing conclusions from it and thus integrating certain theories of practice. This method has the merit of strongly involving young students, but it has its limits in secondary classes or higher education.

In other words, these three forms of research intervention in innovation are complementary and rarely carried out together. A deeper, comparative study of their impact on the same collection of innovations would be interesting to conduct, as Jean-Claude Filloux likes to repeat, "being in research is not the same thing as doing research!"

## 3.4. Innovation in schools in European countries

In 1994–1995, we created the European Observatory of Innovations in Education, within the National Institute of Pedagogical Research. It had five objectives:

1) improve the quality of teaching;

2) study strategies for and new approaches to change;

3) understand the cultural process of emergence and development of innovations;

4) make available to the scientific and educational communities information and knowledge on emerging changes;

5) help with professional practice for teachers and trainers.

This observatory found that not all European countries have the same position with regard to innovation and, sometimes, the term does not exist in their country! However, we tried to identify common tendencies in the understanding of change in teacher practices. They numbered three: first, novelty in regard to the immediate context; then, the question of values out of step with those offered by the institution; finally, the particular unusual change.

The study, carried out in about 10 countries, allowed us to identify intersecting factors which explained the development or otherwise of innovation in schools:

– the degree of centralization or of decentralization of the country. This dimension is involved in the question of a *bottom-up* or *top-down* process, that is, innovation controlled from the center (often linked to reforms) or coming from the margins. In reality, it is not that simple. We can in fact see that countries with a reputation for being centralized tend to decentralize (France, Spain, Greece or Italy) and countries known for being decentralized have a tendency to centralize (United Kingdom, Germany). This simply means that innovation is located between the two: too much decentralization creates a cacophony, and too much centralization leads to denial and paralysis. It is through this middle ground that EU countries are seeking a way forward. It seems that innovations emerging from the field are more or less taken up by the center through official decrees: this is what some call a "spiral" approach to innovation. Nothing prevents the authorities being ahead of what happens on the ground and in this case, it is the central authorities who produce a decree to innovate, which is more or less followed. Innovation is born from this dialectic where the center offers a framework into which innovations are entered, even in Germany, as a compromise. The greatest importance is in the spaces of freedom;

– the role of parents. In Germany, education is part of the competence of parents who play a role in the innovations put into place by the different schools. It is obvious that such a thing may contradict the republican values of school which make parents simple consumers of school;

– the weight of values and of mindsets. That is, the fact that innovations reflect the social culture of the country. Thus, Finland and Denmark defend their strong values of holism in education, where students go to school to develop on the social, artistic and ethical levels. In France or Austria, these values are not part of the consensus, and innovations aiming for total education still have problems finding support in the various layers of the population, even if over the last 20 years some change has occurred. This is even more important in that innovations are based on values;

– the relative weight of private or alternative schools. This may intervene to force public schools to move, assuming that they have a flexible national framework. In fact, private schools, whether contracted to the State or not, have commitments with regard to public norms, though more flexible, and thus can put in place new pedagogies and varying educational structures. In Germany, a private school, to get permission to open, is required to show that it is innovative compared to public schools! In Denmark, private schools follow the framework offered by the State, although this is flexible in its requirements;

– the hierarchy of knowledge or disciplines, as well as their official demarcation. This may play a role in hampering the development of innovations. For example, in France, working on social knowledge is less accepted than working on the teaching of the various disciplines. Similarly, in Greece, classical knowledge is more easily introduced than professional knowledge. In Denmark and Finland, it seems that knowledge of aesthetic and social kinds is promoted;

– the presence of intermediate bodies (such as pedagogic advisors or inspectors who are trained in supporting the continuing development of innovations). This allows us to instill the spirit of innovation among teachers. Similarly, school councils and school leaders, through the restructuring of their role in decentralization in France, allow the encouragement of innovation, on the condition that they are explicitly invited by the official authorities to do so;

– the presence of mechanisms for encouraging innovation. This leads to their development. These mechanisms may be clearly identified Institutes as in Portugal or France, putting in place "rewards" or "prizes" for the best innovators chosen by a jury according to criteria responding to the thematic requirements of policy or previous topics;

– the presence of teaching unions and, above, their role regarding innovations. Some unions even take charge of continuing education (Austria).

CERI[9] published in 2014 a study whose goal was to measure the role of innovation in education compared to other social sectors. They drew from it some very interesting comments which go against the received wisdom that innovation in education is weaker and less frequent than in other social sectors.

Currently, the OECD is working on a project titled "strategy for innovation in education and training" launched by the Ministers of Education at their meeting in May 2007. According to the Oslo Manual, this project defines innovation as "the introduction of a new or substantially improved product, procedure, organization or method of marketing". For this organization, "innovation therefore does not only cover invention, but also

9 CERI, the Center for Research and Innovation in Education, is an agency of the OECD.

improvement", and it distinguishes several degrees of novelty: novelty for the world, for the sector or for the institution (or firm).

This work is based on two elements: the learning organization and human capital in liberal societies. It is less about seeing innovation as a single, isolated and circumscribed thing than taking into account the global movement towards improvement in the light of student results. The secretariat of the OECD has decided:

> "[...] to exclude from its field of reference everything concerning the impact or the shape of reforms decided by governments, to limit itself to analytical description of the ecosystems which allow continuous improvement and make radical innovation possible in the educational sector".

It seems that three ways of understanding the innovating school and innovation[10] coexist in the countries involved in this OECD study:

– the countries that see innovation as an action emerging from the initiative of teachers or of people within a school, of course after the diagnosis of a reality in respect to the objectives. This innovation is noticed and then must integrate, more or less well, with the other activities involved in an already-existing school project. This integration is based on negotiations between the different categories of staff represented on the school's Board of Directors. This activity, identified as such, engages in a process of recognition, evaluation and possible validation over the medium term. This is the situation found in France, Spain or Portugal;

– the countries which believe that innovation does not exist as an isolated activity. It is the school as a whole, as a system experiencing evolution of a continuous nature, evolution which may anticipate the future and be out of step with civil society. For example, phrases are used such as the innovative school in a knowledge society, the latter being not clearly defined. This is the situation found in Germany, Sweden or Finland;

– the countries which consider innovation within schools as the pragmatic result of a *process of finding what works*, that is, innovation which works and is declared to be functional with regard to stated objectives, without any other analysis of the said objectives. This third variety relates more to innovation as seen as an adaptive response to the world as it is. This is the

10 CROS F., *L'agir innovationnel: entre créativité et formation*, De Boeck, Brussels, 2009.

case in Denmark, where innovations are based on matching between elements of civil society (development of information technology, development of cultural and linguistic diversity, etc.). There is thus a dilution of innovation among the set of adaptive responses that a so-called innovating school would use.

In other words, the set of elements discussed leads us to question the training of education staff – not only their initial training, but also their continuing training[11].

## 3.5. The five stages of public policy for supporting innovation in France

Official bodies in France have been taking into account innovation in schools as a source for transformation of the educational system since 1960[12]. This has evolved according to concepts of innovation and its possible role in this transformation. We have been able to identify, over the last 60 years, four ways of managing innovation in public education policy, to which we have added a fifth (the start and end dates of the periods are indicative).

### 3.5.1. *Innovation, a marginal subject (1960–1980)*

We can pinpoint the first time that innovation was considered by the Ministry of Education as cautiously positive. Up until this point, even in the 1960s, it was seen by official bodies as a challenge to the established order, closely linked to the movements of 1968 and the Amiens Symposium which brought together the greatest pedagogues and researchers of the age to envisage a deep-down transformation of the educational system, including support for innovations closely linked to research. This symposium has had little subsequent follow-up.

11 CROS F., *Préparer les enseignants tout au long de la vie: une priorité européenne?*, L'Harmattan, Paris, 2005.

12 For reference, let us stress that, starting in 1957, the United States launched a great number of reforms and innovations in education after their humiliation at the Soviet Union's launch of Sputnik, which was doubtless not without (later) influence on Western educational systems, as in France.

Innovation was then taken up by the pedagogic movements in a spirit of advocacy and of opposition to a structure considered bureaucratic, impeding the possibility of innovating. It is thus on the margins that innovation developed, most often supported by pedagogic movements. This was not without challenging the Ministry of Education, to the extent that this increased the power of the protesters.

### 3.5.2. *Innovation as a process to be channeled (1981–2000)*

Innovation became, starting in 1981 (the rise to power of a left-wing government), an issue for many middle managers who had to respond to the growing demands from teachers for authorization of allowances for innovating. The situation had become unbearable.

Partial decentralization in 1983–1985 had given schools a preview of autonomy, particularly with the creation of the secondary level of the Local Public Establishment of Education (EPLE), to which was added the framework law of 1989 which imposed school projects, extended to all schools in 1990. Innovation was increasingly intertwined in these new schemes, with greater or lesser happiness.

The creation in 1991 of University Institutes of Teacher Training (IUFM) reshuffled the deck, particularly by publishing a syllabus of teacher competences which explicitly included the competence of innovating. However, it did not explain how: it was reduced to a pure command.

Faced with the appearance and increasingly forceful advocacy of innovators, encouraged by middle managers, the Ministry could no longer ignore them, even if innovation was still approached gingerly. After long hesitation, the Ministry proposed in 1995 the creation of a mini service (note the reaction time of the Ministry: no less than 20 years!) comprising a contracted person from the private sector and two seconded teachers[13].

After being attached to the inspectorate, this service was placed under the authority of the Directorate-General of School Education. It was now known

13 To read the account of one of the seconded staff leaders see FAUCQUEUR C., "La valorisation des innovations pédagogiques", *Économie et Management,* no. 119, CNDP, April 2006, http://www2.cndp.fr/revueEcoManagement/pdf/119/068073.pdf.

as the INNOVALO mission at the national level[14]. Innovation was now seen as an object which might be capitalized on through promotion, independently of its authors, as an action whose process could be isolated and might give ideas to other teachers, in a sort of incentive to change their model.

Little by little a national mechanism in network form was built, piloted by a Ministry office with a representative (often an elementary or high school inspector) of this mission in each academy, designated by the Rector of the academy, surrounded by permanent staff such as teachers, pedagogic advisers, heads of school and partly seconded from their usual tasks. This regional team had the job of informing the mission at the local level, of encouraging teachers who were keen to innovate, and of making connections with local research laboratories which might help enlighten them conceptually. This encouragement also consisted of detecting innovations which emerged within the Academy and getting middle managers to monitor their progress.

But all this happened under the control of the Ministry, which determined a policy of innovation according to prioritized themes. Academy leaders benefited from gatherings and training at the national level.

Every two years, these national themes were renewed or reiterated according to governmental direction. This biannual national plan, called the PNI (National Plan of Innovation), set out a list of these themes, to which other themes might be added according to the situation of the academies and their priorities[15].

---

14 Note that this measure arose in 1993, in the context of the 158 proposals of the *Contract for Schools* initiated by the Ministry of François Bayrou, in which explicitly appeared a commitment to promote "innovative" pedagogic practices, bearers of educational success. Measure 53, under the heading "Promoting pedagogic successes", foresaw that "a means of promotion and follow-up of pedagogic successes will be put in place in each academy". Measure 87 linked pedagogic innovation and initial teacher training: "Innovative approaches in continuing training and in adult education will be the object of discussion and reflection in the goal of drawing lessons for initial training". Finally, concerning organization, measure 91 indicated that at the beginning of the 1995 school year, "the description and diffusion of experiments carried out by schools and firms will be ensured by the resource centers established at Boards of Education".

15 For example, the PNI of 2003–2005 included the following themes: new ways of organization for learning; diversity of intelligences and plurality of paths; student expression in their learning and at their school; innovation in evaluation; development of collective work in the school and with its external partners.

It is on this basis that appeals for submissions were set up, for teachers to come forward with those innovations of theirs which came under the scope of the appeal. Teachers could then submit individually or collectively. They had to inform their director, who would often not be opposed as long as the resulting disruption was not too great for the rest of the institution.

Thus, innovators had to directly deal with the local branch of INNOVALO and engage with those who "filtered" innovations according to criteria linked to the themes and the feasibility of the innovation. In fact, acceptance of the innovation by academic bodies was accompanied by possible allowances including timetable changes, leave for innovators, or even overtime. This distribution depended on the financial generosity of the Ministry in its funding of the Board of Education.

Once this balance had been struck, innovators were required to participate in regional or national demonstrations of promotion of their work. They also had the obligation to fill out standardized forms to be registered in the regional and national innovation file, according to categories linked to the level of students involved, to themes and to other categories such as pedagogic practices, etc. The national file allowed the diffusion on paper of these innovations and their registration on the Internet sites of Regional Centers for Pedagogic Documentation (now registered on the Canopé network).

The function of this mission lasted over four PNIs, the first in 1995, the last overlapping with the creation of a National Council for Innovation and called the *Plan for Encouraging Innovation*. The network structure allowed the establishment of personalized support for those innovations accepted (often called "innovating projects") according to an Academic Innovation Plan (PAI) with a writing contract for innovators, so as to create content for the regional and national capitalization sites.

### 3.5.3. *Innovation becomes a competence for actors in the education system (2001–2010)*

Successively, three waves of syllabi of teacher competences were also published (2006, 2013 and 2015) which supported, each time more strongly, the competence of innovation in teachers.

But things were not so clear. It became more and more obvious that innovation posed problems concerning allowances, limits to change, and the autonomy of public service staff who owed obedience to the Ministry; the local level (that of the innovations) could not be clearly defined in application.

The willingness of official bodies to consider innovation in schools was increasingly clear and resulted in October 2000 in the creation, under Minister Jack Lang, of a National Council of Innovation for School Success (CNIRS) comprising people involved in the Pedagogy and School Innovation Research Service, whose goal was to identify, support and evaluate innovative actions. It was also based on the need to coordinate, in consultation within the Ministry of National Education, a set of practices, in order to develop and generalize the most effective innovations.

This council allowed the creation of 10 experimental institutions, of which two remain. Luc Ferry, the new Minister of Education, abolished this Council[16] arguing that the Directorate-General of School Education (DGESCO) should take over its mission.

During the short life of this council, the INNOVALO mission continued to function in parallel with the same activities that it was previously entrusted with. The leader of this mission could attend meetings of the Council without, however, there being a clear vision of the responsibilities of each of these bodies. This mission continued to function after the end of the Council, in a less structured manner.

This multiplication of acronyms has doubtless not helped their visibility. However, we must stress that the Ministry took its role seriously and set up training meetings for these regional leaders, as well as events on national days.

The year 2005 saw the creation within DGESCO of the Department of Research and Development, Innovation and Experimentation (DRIE), whose dissemination was based on the old academic leaders in charge of supporting innovation, that is, the Academic Advisors in Research and Development, Innovation and Experimentation, under the acronym CARDIE. In reality, many CARDIE personnel had previously belonged to the decentralized network of the Innovation mission of the Ministry.

16 In reality, he beheaded it by abolishing its secretariat, provoking the resignation of most of its members.

Parallel to the creation of the DRIE within the Ministry, the publication of Article 34 of the Framework and Schedule Law for Schools of 2005 (BO no. 18 of 5 May 2005) would reframe innovations, now called experiments, within a system which was at the same time wider and more constraining. It said:

> "In each school and every institution of public education, a school or institutional project is developed with representatives of the educational community. The project is adopted, for a term of between three and five years, by the school council of the Board of Directors, on the recommendation of the teaching team of the school or of the Teaching Council of the institution as it concerns its pedagogic section.
>
> The school or institutional project defines the specific modalities of enacting national objectives and programs, and specifies the school and extracurricular activities that contribute to them. It specifies the ways and means put into place to ensure the success of all students and to involve parents to this end. It also determines the means of evaluation of the results achieved.
>
> Subject to the prior authorization of the academic authorities, the school or establishment project may provide for carrying out experiments, for a maximum period of five years, relating to the teaching of disciplines, interdisciplinarity, the pedagogic organization of the class, the school or the institution, cooperation with partners of the educational system and exchanges or twinning with foreign educational institutions. These experiments will be subject to annual evaluation.
>
> The High Council of Education every year issues a balance sheet of experiments carried out in application of this Article".

This involved two major changes with regard to innovation as envisaged previously: on the one hand, innovation was no longer called innovation, but "experimentation", which slightly shifted its meaning, more towards a controllable protocol and project; on the other hand, making every innovation fit into the straitjacket of a school project with the agreement of all parties strongly eroded its original and transgressive aspect. In addition to running this gauntlet, we must add the obligation to deposit a file with the Board of Education which evaluated the relevance of the request, could refuse it and, if they accepted it, change its direction. The decision of the Rector was final. In short, some major characteristics of innovation as we have discussed it were significantly eroded!

There was, however, a benefit in including innovations in the institutional project, in relation to the coherence and cohesion of the teaching body and parents. This benefit might reveal itself in a situation of strong school autonomy with a choice of teaching teams, greater liberty in pedagogic practices and structures, in a spirit of consultation between staff. This ministerial desire for successive binding stages contributed to the stifling of innovation, even if at the same time they opened the Expérithèque to capitalizing on innovations, on the Ministerial site and on the academic sites.

A team of researchers were put in charge of evaluating these experiments[17]. After consultation covering many documents, meetings, interviews and observations, this team made three main remarks which clarified official ideas concerning innovation in schools:

– the academies develop this function in a very diverse manner (some pursue many experiments and others very few);

– the themes of innovations are very diverse (the themes prioritized by the Ministry seem less evident);

– some academies make a distinction between innovation and experimentation while preventing the possibility of complete inclusion in the institutional project.

The fragmentation of these relatively reduced and ephemeral structures makes these innovations/experiments very dispersed and reduced in their local reach. In addition, the weight of written work and mandatory evaluation, often linked to institutional criteria which have little to do with the actual specificity of the innovation, have discouraged more than one from going down this path.

Researchers found it difficult to assess or even to count the number of innovations/experiments considered by these bodies. In other words, the Ministry claims to be interested in innovation, but does not use all its means

17 REUTER Y. (ed.), Rapport sur les expérimentations liées à l'article 34 de la loi d'orientation et de programme pour l'école de 2005, submitted to the High Council of Education. June 2011.

to promote and structure its effects. However, incontestably, the Ministry has taken on board the idea that innovation is part of the development of the system, of the school landscape and of demonstrating new professional competences, while still keeping a firm grip on any possible excesses.

The idea of including innovations in the school project doubtless implied a conception of change in the educational system through an idea of innovating schools, where all actors are associated and involved with a progression thought of as irreversible. A combination of *top-down* and *bottom-up* is on display here, although no doubt forgetting that innovation is not only an action introduced, it also brings into question the epistemological and philosophical basis of schools.

### 3.5.4. *Innovation as a social regulator (2011–2017)*

This period is marked by the creation in April 2013, a decade after the previous one, of a National Council of Innovation for Educational Success (CNIRÉ), under the Peillon ministry; its term of office was four years. Its mission statement is as follows:

> "– Promote the spirit of innovation in regard to success in schools and in educational success.
>
> – Propose directions in innovation policy in the field of school and educational success. Its field of competence extends to teaching at both primary and secondary level.
>
> – Appraise and evaluate innovative practices enacted in the territories.
>
> – Diffuse the most relevant practices.
>
> – Organize the debate on innovation with actors in innovation (leaders of the educational system, researchers, representatives of associations, etc.).
>
> – Bring to life, in association with DGESCO which oversees it, the network of academic advisors in research and development, innovation and experimentation (CARDIE)".

Moreover, the DRIE has developed different tools of communication such as *The Letter of Innovation* or the creation of a new database, *Expérithèque*, replacing the previous database, and the creation of a professional network of innovation on the model of social networks (RESPIRE).

The number of databases available to the teaching and education public has multiplied over the last few years, whether with regard to discipline, to general themes or in expansion to educational staff. For example, Viaéduc is a network of educational professionals in the form of a public interest group (GIP), on a collaborative platform run by the Ministry.

Figure 3.2 translates the DRIE's intentions.

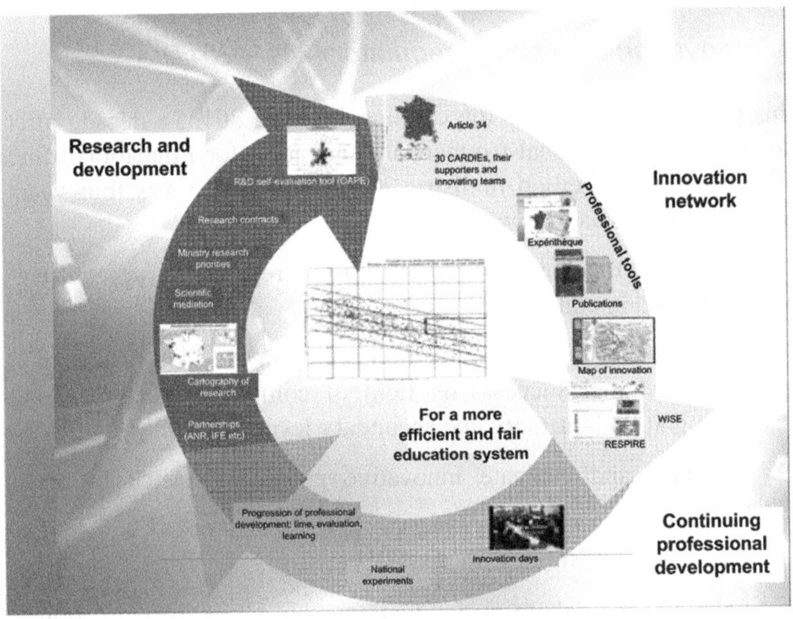


**Figure 3.2.** *Schema of the functioning of a national policy of innovation in schools*

The DRIE has thus observed four types of experiment, as presented in Table 3.2.

| Structural experiments – type 1 | Pedagogic experiments – type 3 |
|---|---|
| Concerted collective organization<br>A sustained and supported team<br>Widens the classic framework to the level of the institution (allowance)<br>**Internal/external regulatory evaluation** | Small team<br>Sharing in an action<br>Touches lightly on the organizational frameworks of a discipline or a class (allowance)<br>**Evaluation must be supported** |
| **Institutional innovations – type 2** | **Pedagogic innovations – type 4** |
| New or recent collective dynamic<br>Involves several levels (discipline, levels, institutions)<br>Does not touch the classical framework<br>**Evaluation not always organized** | One or more persons involved<br>Introduction of a new element into the framework of teaching, of a function, of a class or of a level<br>**Does not significantly disturb organization (time, spaces, evaluations)** |

**Table 3.2.** *Typology of experiments identified by the DRIE*

As for CNIRÉ, it meets twice a year and is composed of 39 members. It will not have escaped anyone that its name is not identical to that of its predecessor council, particularly concerning the qualification of the objective. It is no longer a matter of "school" success but of "educational" success, that is, involving a wider educational landscape, which offers a wider spectrum of participants, more often appointed for their leadership in Ministries or organizations which are closely or not so closely involved with education.

For the first two years, this council was led by a university sociology professor, and for the last two years by a teacher known for her vibrant participation in pedagogic movements. Its composition was enriched by actors in education on this occasion. Set up to last for 4 years, its mission ended in 2017. It has produced two reports which may be consulted on the Ministry's website, whose content attests to the liveliness of the proposals and themes which were worked on and may be deepened[18].

18 The first report came out under the name of D. Lapeyronnie (10 November 2014) and makes 25 proposals for change; the second report came out under the name of P. Watrelot (29 March 2017) and makes 10 proposals.

Besides this Committee, in 2015, the Minister of Education, Najat Vallaud-Belkacem, announced that the links between innovation in schools and research had to be strengthened. To this end, she entrusted a mission concerning the Carnot Institutes[19] of Education to Roger Fougères, who submitted his report in July 2016 after experimentation at the regional level in Auvergne-Rhône-Alpes. These Institutes are still under construction and thus it concerned testing the relevance of this alliance between innovation and research in education. This mission was hosted by the Institute for Training and Education (IFE) in Lyon and funded by the Ministry. This mission had three objectives:

> "First, to develop pedagogical innovations backed by research through a *bottom-up* type of approach, starting with consulting and questioning teaching teams and putting them into direct contact with teams of researchers to co-build projects; then, to promote new research activities which could qualify as 'engaged', on subjects which are today badly understood by research, which are real 'locks' for innovation; finally publishing their works while prioritizing initial and continuing training for the students of ESPEs, active teachers or even academic managers"[20].

All this is part of the Plan for Investment in the Future (PIA 3). The experimentation carried out in the Auvergne-Rhône-Alpes region, following the setting up of calls for projects stemming from a link-up between teachers and researchers, seems interesting, but it is still too early to measure its real impact on the development and strengthening of innovation in schools.

Note that originally, the Carnot Institutes (CIs) date from 2006 and were developed in industry around central prescribed themes, as the Carnot

19 The Carnot Institutes, born in 2006, are structures for bringing together universities, the "grandes écoles" and businesses. The goal is to develop the competitiveness of firms through the creation of new products sold on the market, thus delivering substantial profits. The question is still that of wanting to put innovation in schools under the same model as technological or product innovation. It seems that the question of the different values associated with schools is underestimated here.

20 FOUGERES R., Final report, mission Institut Carnot pour l'éducation, définition, expérimentation, essaimage, IFE, 2016.

Institutes would want to do in education. In the ICs dealing with business, every IC is devoted to a theme or a field of professional intervention, such as the Brain Institute, the Institute of Mines, the Institute of Mechanical Industries or the Institute of Public Construction and Works.

Finally, to this collection of measures trying to manage innovation in schools, we can add the mission of François Taddéi. The latter, the director of research at the Center for Interdisciplinary Research founded in 2015, was put in charge in September 2016 of a mission dedicated to pedagogic innovation. It particularly aims to[21]:

> "– Promote within the educational system a logic of trust conducive to the development of pedagogic innovations backed by research.
>
> – Stimulate research on education to expand knowledge.
>
> – Develop the diffusion of knowledge for the benefit of initial and continuing training".

On the site, it is added:

> "For this, the mission will have to propose new modes of organization and coordination, priorities for concrete actions and perspectives on mid-term evolution. An intermediate report is expected before the end of 2016. It will be a contribution to the great choices which must guide the specifications of future calls for projects under PIA 3, and will launch a stage of prefiguration of a new means of organizing and piloting research and innovation policies in our educational system. The consolidated report is expected for March 2017".

The report did indeed come out in March 2017, and it concluded with 10 proposals which seem to us to be important to mention here, as they herald a new political way of conceiving innovation in schools:

– more support for research in the initial training of all actors;

21 Extracts from the website of the Ministry of Education, May 2017, http://www.education.gouv.fr/cid115649/vers-une-societe-apprenante-rapport-sur-la-recherche-et-developpement-de-l-education-tout-au-long-de-la-vie.htm.

– encouraging the engagement of teams in experiments and their networking;

– intensifying research on education;

– managing physical third spaces;

– setting up digital third spaces;

– promoting a handbook for lifelong learners;

– promoting norms, methods and frameworks for trust;

– conceiving and promoting ethical rules;

– territorial alliances (of learning territories);

– prefiguring a research alliance.

These different missions try to introduce a new spirit involving all the bodies involved, closely or otherwise, in education (which, if we take the thing at its word, involves all of society) in a continuous process of learning.

Note that François Hollande's last Minister of Education created multiple missions for reflection and proposals dealing with the development of innovation as steering of the educational system, in particular possible investment in research on education *(evidence-based policy)*, thus marking an international orientation without raising the question of knowing whether, by itself, research can objectively solve all the questions raised in schools.

But the advisors are not the funders, and it is difficult, through these ambitious proposals, to detect a process that will allow them to introduce this spirit[22].

There is a kind of gap between these proposals and the actual management of the educational system whose job it would be to define scalable procedures, to move from the current centralized system to a national or even an international system, including the local in a wider globalization linked to research. In these conditions, innovation becomes a

22 Note, however, that the Council headed by a secondary teacher took greater interest in what is done in schools than what should be done, which the leader called "changes on the quiet".

component of social evolution which is itself learning, and disappears in its distinctive, transgressive form. It becomes the very definition of the functioning of the educational system and its evolution. This is without doubt the fate of all social innovation.

The whole question rests on the articulation between pedagogic initiatives and the determinants of a policy decided (in consultation) at the national level. The strategy of links between the two levels rarely supports reforms, which most often fail for three reasons:

– the first, often mentioned: hardly is a reform launched, when another reform replaces it, without it even having benefited from evaluation allowing its effects to be measured;

– the second reason is time: that of the school (and also that of research) is not that of politicians. Every change in the school system requires a period of adaptation and stabilization;

– the third reason is to be found in the adjustment of innovations. If we consider that every teacher is driven to innovate from the very fact that no professional action can be reproduced exactly (incremental minor innovation considered as a professional competence when it is reflected and worked on), innovation which might help change the educational system lies at another level of innovation. It is thus a question of a level which breaks more with habitual or organizational practices while not being revolutionary, with a slow implementation accepted by others, whether by going in the same direction, or by imagining something else without causing a contradiction. This tension between center and periphery is a matter of articulation, where reforms are based on innovations which have already been carried out and encourage their progressive generalization, without such actions as clumsy and obstructive decrees which see opponents as idiots.

In fact, every opponent of an innovation has their reasons which we must seek to understand and discuss in a curriculum of continuing training. The latter is an obligatory part of a teacher's career through a career-long professional project, and leads to change which is accepted, even enriched. The structural, social and economic context is a level which may sometimes be the determining factor in changes in behaviors, mentalities and attitudes.

What might we then say about these four periods of the political management of innovation in France (to which we have added a future period)?

First, we see a change in the attitude of official bodies which, fearing excess and troublemaking from the innovator, invite the latter to be one of the leaders of change in schools.

Subsequently, innovation is no longer conceived in the same way: from an object which can be isolated from its authors, it is seen as consubstantial with the changing practice of the teaching profession, even the instrument that allows us to evaluate the professional competences of teachers.

Finally, from a position of rejecting all reforming activity through innovation, official bodies have gradually found it to be an ally in the evolution of the educational system through its adaptive and mobilizing qualities and in phase with a new society based on mobility, individuation and equity.

In Table 3.3, we have intentionally added a future phase which purports to portray the role of innovation over the next 5 years.

| | **Model I**<br>**1960–1980**<br>**Innovation as an object** | **Model II**<br>**1981–2000**<br>**Innovation as a process** | **Model III**<br>**2001–2010**<br>**Innovation as a competence** | **Model IV**<br>**2011–2017**<br>**Innovation as a social regulator** | **Model V**<br>**2018–2022**<br>**Innovation as a survival need for any learning social system** |
|---|---|---|---|---|---|
| **Historic context** | The State claims a monopoly over innovation which modernizes and structures school on condition of abandoning part of its transgressivity | Innovation is initiated by the State as a necessary safety valve for the system | Innovation is a lever of change in the system, particularly in professional practices and in training | Innovation is the motor of development of all institutions as a lever of change | Innovation must be encouraged and stimulated as an integral part of any effective policy |

| | | | | | |
|---|---|---|---|---|---|
| **Nature of innovation** | A product to be distributed after hierarchical approval | A process to be controlled, but a possible solution to certain problems | A way of integration into the system and an indispensable professional competence | A major ingredient of evolution and adaptation for any system, on condition that it responds to major social priorities | A modality consubstantial with any educational system according to rules linked to its performance and to ethical and national obligations. It is part of an organization and allows its structuring. It is also specific to neoliberalism in schools (rise in power of alternative schools) |
| **Dominant level of analysis** | Macro (the whole centralized system) | Meso (institution and networks) | Micro (teacher or teaching team) | Network inside and outside the system | Organizational level of a school or even of a network of schools |
| **Origin of innovation** | Originating outside the school or in intersecting marginality | Mixed origin and endogenous development | Endogenous origin | Collective exogenous and endogenous origin | Originating through successive outbreaks and many incentives (sometimes a requirement for innovation) |
| **Diffusion of the innovation** | Planned diffusion from the center to the periphery through official channels | Central monitoring, but spread from one periphery to another periphery | No real diffusion: transfer encouraged locally | Locally negotiated creativity and national diffusion by the center | Installation of a central and regional mechanism for innovation and incentives inspired by existing innovations (demonstrations and regional and national meetings). Massive use of ICT, development in networks |

| | | | | | |
|---|---|---|---|---|---|
| **Form of change** | Sequencing of innovations legitimized by the central power | Selection and adaptation of innovations retained as exemplary (local reappropriation) | Incremental change through osmosis and professional development of innovative capacities | New collective constructions in which innovations are registered | Innovations are the center of change and reform is based on the process of innovation. Organizational aspect of change in schools |
| **Role and status of innovators** | Tolerated marginal actor | Actor as bearer of the innovation (negotiations, evaluation, power relations) | Author of innovations in recognition of agency | Creator of new practices which bring improvement to the system | Collectives which innovate or create a new school (autonomous research in the school) |
| **Type of evaluation** | Judgmental assessment led by the innovators and their goal | Formative evaluation allowing the identification of conditions for its implementation | Evaluation of the aspect of promotion of the novelty and of its effects | Collective and supportive evaluation by middle managers | Evaluation of the effects of innovations on the school as an organization with expansion to external partners |
| **Type of support** | Orthopedic | Controlled diffusion | A social process to promote | Actors taking responsibility in accountability and empowerment | Response by the director of the school in an organization of innovations as the identity of the school to be defended in its pedagogic choices and values in line with research |

**Table 3.3.** *The five stages of political and social management of innovation in schools in France from 1960 to 2022*

In an article[23] in 2009, concerning a study of the obstacles to introducing change, Sylvie Trosa identifies four types of behaviors (ideal types) which may influence us, faced with this change:

– *the heralds of innovation* focusing on the level of micro-change which, paradoxically, facilitates the sustainability of the system because the micro never becomes macro. They demonstrate their innovation limited to one field and one context: they are zealous and supported by the administration under the benevolent gaze of intermediary official bodies. They endorse the idea of an administration committed to the modernization of the school. These innovations of limited scope become exhausted, because quantified evaluations were not built to take them into consideration. They refuse to direct their capacity for innovation towards more general formalization. They require investment in time and staff which is sometimes unreasonable. The question which is often evaded is that of the change of scale of innovation and of questioning these undeniable capacities for positive self-adjustment and boldness. This is not an apparent change, as it supposes a new culture which accepts the validity of the fact that there is no one right answer, but it is limited in its scope;

– *systemists specializing in systems analysis*, which is based on the interplay of actors in a complexity which sometimes approaches paralysis. All change must be preceded by a precise study of the relations of power between actors and be based on behavior of limited rationality in their actions. They argue the necessity of dialog with unions, a thorough preparation of minds at all levels of the hierarchy. The difficulty comes from the contradiction between the time needed to carry out this vast study and the successful arrival of ministers who want to leave their mark on the educational system. We thus return to the question which has been raised many times in this work, that of political and administrative continuity at the highest level of the hierarchy, without forgetting the determining financial commitments. To this we add the lobbying from certain parts of civil society who prefer to maintain a certain elitism. Remember the innovation concerning the dual role, a few years ago, of teachers in junior high schools, innovation which was tested and refused;

23 TROSA S., "Les réformes vues d'en bas", *Pyramides*, no. 17, 2009, published online 7 December 2011, accessed 14 May 2017, see http://pyramides.revues.org/636; "L'État est-il encore régalien ou devient-il un réseau social?", *Pyramides*, no. 25, 2013, published online 10 March 2015, accessed 13 May 2017, see http://pyramides.revues.org/967.

– *producers of reports or studies*, consisting of gathering the best minds in the country and getting them to produce proposals, supported by some examples of innovation, whether in France or having outright scoured the globe through countries considered at the cutting edge of change in schools and which are exceptions. Thus, everyone falls into agreement on their general consensual proposals which, moreover, reprise some previous reports by putting them into the style of the current context. The problem is that these proposals do not enter the realm of the concrete, which is where problems arise, of knowing how to do it, according to which process, and the constitution of an operational strategic team. In other words, consensus is only established when one remains vague and only writes what the reports allow;

– *partisans of French exceptionalism* who insist on the fact that France is a country culturally founded on a multiplicity of rules. These culturalist approaches praise our administration without yet asking themselves about possible actions that would allow the educational system to evolve in the desired direction.

Table 3.4 sums up these four attitudes.

| | |
|---|---|
| ***Innovators through "reports"***<br>– Operationalization of the change is lacking<br>– Not put into action and monitored | ***Believers in French cultural exceptionalism***<br>– The argument is unbeatable, since any proposed solution is rejected as contrary to French culture |
| ***Recognized innovators***<br>– They propose concrete changes, even if limited in scope<br>– But due to their isolation (even with the best will of networks or symposia or periodic continuing training activities), the non-generalization of these innovations is not taken into consideration and causes the blockage of relevant and systemic channels | ***The systemists***<br>– They have a fine understanding of interrelationships between the problems and the levels on which it is necessary to act simultaneously<br>– But this raises questions which are so weighty as to appear insurmountable |

**Table 3.4.** *Four attitudes of innovators*

Refusal of change is often included in the concept of resistance, which explains nothing. This policy of development of innovation lacks an enduring structure of governance, taking things one after the other, in a desire to continue over several decades, avoiding contradictory orders and orientations in their realization.

The principal dimensions are based on delegation of power and school autonomy[24] (budget, recruitment of staff including teachers, choice of curriculum contents in accordance with a wide national framework of acquisition of competences on a common base, organized and systematic sharing of innovations, hierarchical relationships directed towards help and support as much as control, etc.). This is on the condition of a contract of objectives and an essential partnership between the pedagogic and administrative fields.

The paradox comes from the fact that this concerns a centrally driven policy while the freedom of action of teachers, although relative, has brought us a great richness of innovations, which remain on the field of their emergence, without being solidly and continuously taken up, which Pavé calls "the silent modernization of public services"[25].

At a conference in Istanbul, following an experiment in developing a new school curriculum, Hellström[26] identified 20 conditions for the development of innovation on a wide scale:

1) create a stable, clear, comprehensible national *curriculum* which does not leave schools to muddle around in confusion;

2) do not just give instructions, but leave room for discussion;

3) create opportunities for cooperation;

4) treat teachers with respect, even if some are "professionals" and others only "practitioners";

5) leave schools to make choices;

6) give schools questions to which they can respond, and not just work to do;

7) schools can make a difference: find a balance between pressure and aid;

---

24 On this subject, see FEYFANT A., "A la recherche de l'autonomie des établissements", *Dossier de veille*, no. 118, 2017.

25 PAVE F., *La modernisation silencieuse des services publics*, L'Harmattan, Paris, 2006.

26 HELLSTRÖM M., "School curriculum – is it worth trying?", in SAHLBERG P. (ed.), *International Conference on Curriculum Reform and Implementation in the 21st Century*, Istanbul, 2005.

8) need for paid time;

9) organize cooperation between schools and administrations;

10) make school directors into pedagogic leaders;

11) use pilot projects;

12) listen to students;

13) allow access to necessary resources (time and money);

14) give schools a series of successive goals, attainable in reasonable time;

15) advise schools on ways of organizing curricular work;

16) create ways of rewarding difficult work;

17) take care of teachers and their well-being;

18) give development time to schools;

19) get involved in curricular work;

20) train institutions in how to identify problems and overcome them.

Taken as a set, these conditions seem to cover well the recommendations which have been made over several years. The real question remains that of "what do we do, concretely, to put them into operation?"

# Conclusion

The term "innovation" has never been so widely sought after than over the last 50 years, by any and all tendencies, whatever their philosophical or ethical horizons.

In other words, this word appears to cover positive impulses leading towards a radiant future. However, certain contemporaries, in the light of different uses of the term and its concrete applications, have not seen any beneficial values added in it. They therefore back away from this suitcase of a word, embedded in a mist from which it obviously does not want to emerge.

Basically, rather than looking at the word, is it not better to analyze the intentions of those who utter it and the effects they aim to produce?

In fact, the word has not had this positive notoriety throughout our history. Already, Montaigne in his *Essays* spoke of *novelty* to vilify it and to underline that there was nothing worse than a new thing which imperils institutions, which are difficult to maintain as they are.

If the word innovation fits so well in our time, it is because it is strictly linked to the ideas of the time. Now, few experts have spoken of the relationship between innovation and time, or more precisely, of the representation and the execution of activities in time. This idea of time contains two dimensions: that of speed seen from the linear angle and that of the future.

Speed defines our era which is always rushing, abolishing the distance and duration of human activities to place them in immediacy (mobility, courier post, high-speed trains, the Internet, etc.): everything must be here and now. For innovation in schools, it is the same thing, which sinks any project for change in the educational system that requires time.

This has not always been the case throughout the centuries in France; sometimes the past is considered to be of high value, sometimes the future is glorified. The rise in power for the demand for innovation is unconsciously linked to a "relationship with time which is today somewhat disoriented, our historical compass having jammed[1]". Undoubtedly, innovation in all areas develops according to:

> "a presentist regime characterized by the overriding importance of a present which tends to become the sole and unique temporal dimension, the past no longer providing lessons and the future barely allowing projections, the extinction of the glorious tomorrow contributing, as much as the ebb of religion in the West, to emptying the future of all substance: it is a present by default"[2].

This notion of time is of great importance for innovation in schools as it involves the long-term with regards to change in the school system. This dimension is not yet taken into account. The desire of politicians to want to renew everything when they arrive, before even having taken the measure of the rhythms of learning and developing skills, explains the extent to which schools are sick due to not being understood in their specificity and in the spaces of pedagogic autonomy of teaching teams that need to respond to the needs felt by their students. Finally, innovation finds its niche in the link between needs detected by the teachers, the needs of students and their skills at responding in an original and efficient manner.

Throughout this work, we have wanted to survey the questions raised by innovation in schools in its complexity and its various dimensions. The discourse on innovation often remains superficial, whether due to an activist

1 On this subject, see the following two works: KOSELECK R., *Le futur passé. Contribution à la sémantique des temps historiques*, EHESS, Paris, 1990; POMIAN K., *L'ordre du temps*, Gallimard, Paris, 1984.
2 ZEKIAN S., "François Hartog et les temps de l'histoire", *Esprit*, no. 11, p. 193–199, 2003.

desire for mobility while still believing oneself to be effective, or due to a desire not to move from waiting for the next train… which never comes!

If innovation is part of a socio-historical and philosophical context, strictly linked with a country's religion and culture, it may nevertheless be analyzed *a posteriori* in its execution, its transfer, its diffusion and its evaluation. Different models of understanding these processes allow us to act, while not denying the complexity and the unexpected consequences of human actions.

Innovation in schools appears to us to be a particular kind of social innovation which, if it responds to goals which the institution has difficulty taking seriously, holds at its deepest level the challenge of the future of a generation of human beings who will create the future of society. Now, the second element is essential in its formulation: what kind of citizen of tomorrow do we want to shape? The answer is currently under stress from various directions, sometimes paradoxical and yet vital.

In fact, schools and education are central functions which go beyond those provided by a "public service" such as health or electricity supply. Article L 141-1 of the Education Code recalls the Constitution of France in stating that "the organization of public, free and secular education at all levels is a duty of the State".

Undoubtedly, the State plays a central role, but this should not prevent it from modifying its behavior when facing this guiding role.

This new method of guidance would consist of seeing educational staff as innovators, to whom one gives power and avoids seeing them as performers. In France, we still experience a mixed form of guidance, blending bureaucratic guidance through circulars and orders from above, with experimental efforts at change where we find the experimental institutions and innovation missions. This contradictory injunction prevents any development of creative autonomy among teachers, who are caught between a rock and a hard place.

However, it would not do to forget that innovators find themselves in a globalized context whose influence is growing. The trend, which still exists throughout global organizations, has been to promote *best practices*. However, this tendency has quickly reached the limits of fixed practices, of (supposed) imitation of the sector recognized as innovative. The *one best way*, or the unique solution for all countries, has confronted the complexity of real situations head-on.

However, it is impossible to not see that we are in an open world of people who circulate, holding differing values. The response can be neither a universal solution nor can it hold the specificity of everyone-for-themselves. This diversity demands, on the contrary, political clarity on the values which are to be defended, while not rejecting the values of other countries.

The crisis that we are currently living through regarding schools is based on this absence of consensus, not only regarding the grand principles where we can detect convergences, but also on the paths to be followed to put them into action, where innovations have a central role to play. Gramsci wrote[3]: "the crisis consists precisely in the fact that the old is dying and the new cannot be born: during this interregnum, a vast array of morbid symptoms are to be seen."

Such pessimism can be overcome, on the condition that we take into account the social dimension of change. Since Crozier, we all know that society cannot be changed by decree[4], even if most teachers speak of the urgency of "real" reform. In fact, few believe it to be possible, battered by the avalanche of reforms which have descended over the last 50 years.

The same arguments are always put forward: the unions, bureaucratic inertia, the short-term nature of taken measures and their marginal character. However, there is something more serious: there is no agreement on the nature of the problems, on their scope or on the procedure for their resolution.

If a framework is necessary at the national level to channel initiatives, it determines neither the practices nor the ideas. Innovations are the proof of the skills of teachers (even if some innovations may come from students, though this is very rare!). The single rhetorical and ideological register, a sort of incantation, even if it maintains a certain national consensus, will one day or another have to combine with the concrete, effective actions of teachers, that is with innovations.

---

3 GRAMSCI A., *Prison Notebooks*, edited by R. PARIS, Gallimard, Paris, 1983.
4 CROZIER M., *On ne change pas la société par décret,* Fayard, Paris, 1979.

In other words, innovation, though it may appear to some like a breath of oxygen which allows the old system to continue, becomes something inescapable in the management and development of educational systems which face the society of the future. Innovation is therefore not a gimmick, but a real and growing need for the years to come.

# General Bibliography


AJZEN I., FISHBEIN M., *Understanding Attitudes and Predicting Social Behavior*, Prentice-Hall, Englewood Cliffs, New Jersey, 1980.

AKRICH M., CALLON M., LATOUR B., "A quoi tient le succès des innovations? 1) L'art de l'intéressement, 2) Le choix des porte-paroles", *Gérer et comprendre, Les Annales des Mines*, pp. 4–17 and 14–29, 1988.

ANZIEU D., *Le corps de l'œuvre*, NRF, Paris, 1981.

ARIES P., *L'enfant et la vie familiale sous l'Ancien Régime*, Le Seuil, Paris, 1960.

ARON R., *Dix-huit leçons sur la société industrielle*, Gallimard, Paris, 1962.

AUSTIN J.L., *Quand dire c'est faire*, Le Seuil, Paris, 1991.

BACHELARD G., *L'intuition de l'instant*, Stock, Paris, 1932.

BÉNABOU R., TICCHI D., VINDIGNI A., Religion and innovation, NBER Working Papers, no. 21052, March 2015a.

BÉNABOU R., TICCHI D., VINDIGNI A., Forbidden fruits: the political economy of science, religion and growth, NBER Working Papers, no. 21105, April 2015b.

BENBASAT I., BARKI H., "Quo vadis TAM", *Journal de l'Association pour les systèmes d'information*, vol. 8, no. 4, pp. 211–218, 2007.

BERLINGUER L., *Réinventer l'école. Une école de qualité pour tous et pour chacun*, Fabert, Paris, 2017.

BERNOUX P., *Sociologie du changement*, Le Seuil, Paris, 2009.

BLANCHOT M., *L'Espace Littéraire*, Gallimard, Paris, 1955.

BOURDIEU P., *Ce que parler veut dire. L'économie des échanges linguistiques*, Fayard, Paris, 1982.

BOUTINET J.P., *L'anthropologie du projet*, PUF, Paris, 1990.

BOUVIER A., *Vers des établissements scolaires apprenants. Perspectives pour la conduite du changement*, Canopé, Paris, 2014.

BOUVIER A., *Pour le management pédagogique*, Berger-Levrault, Paris, 2017.

BRASLAVSKY C., ABDOULAYE A., PATIÑO M.L., *Développement curriculaire et "bonne pratique" en éducation*, BIE-UNESCO, Paris, 2003.

CARRE P., CASPARD P., *Traité des sciences et technique de la formation*, Dunod, Paris, 2011.

CHILDERS T.L., CARR C.L., PECK J. *et al.*, "Hedonic and utilitarian motivations for online retail shopping behavior", *Journal of Retailing*, vol. 77, no. 4, pp. 511–535, 2001.

CHILDERS *et al.*, "Hedonic and utilitarian motivations for online retail shopping behavior", *Journal of Retailing*, pp. 511–535, 2002.

CROS F., "Émergence et installation de l'innovation scolaire : pertinence de la théorie de la "traduction", *Transformer l'école*, De Boeck, Brussels, 2004a.

CROS F., "La recherche en éducation : une nouvelle forme d'accompagnement de l'innovation, entre acteurs et décideurs politiques", *Accompagner les réformes et les innovations en éducation*, L'Harmattan, Paris, 2004b.

CROS F., *Préparer les enseignants tout au long de la vie: une priorité européenne?*, L'Harmattan, Paris, 2005.

CROS F., *Écrire sur sa pratique pour développer des compétences professionnelles*, L'Harmattan, Paris, 2008.

CROS F. (ed.), *L'agir innovationnel: entre créativité et formation*, De Boeck, Brussels, 2009.

CROS F., "La diffusion des innovations comme expériences", in MORISSE M., LAFORTUNE L., CROS F. (eds), *Se professionnaliser par l'écriture. Quels accompagnements?*, Quebec University Press, Montreal, 2011.

CROS F., MEYER C. (eds), *La communication interpersonnelle en formation*, Chronique sociale, Lyon, 2014.

CROZIER M., *On ne change pas la société par décret,* Fayard, Paris, 1979.

DAMASIO A.R., *L'erreur de Descartes: la raison dans l'émotion*, Odile Jacob, Paris, 1995.

DANDURAND L., "Autour du concept d'innovation sociale, approche historique et comparative", *Revue française d'Administration publique*, vol. 3, no. 115, 2005.

DAVIS *et al.*, "Perceived Usefulness, Perceived Ease of Use and User Acceptance of Information Technology", *MIS Quaterly*, vol. 13, no. 3, pp. 319–340, 1989.

DEHAENE S., *Apprendre à lire. Des sciences cognitives à la salle de classe*, Odile Jacob, Paris, 2011.

DELVAUX B., Une toute autre école, Pensées libres, GIRSEL-UCL, Louvain-la-Neuve, 2015.

DURKHEIM E., *L'Évolution pédagogique en France*, PUF, Paris, 1938.

ELANSARI M., Afrique: Transition éducative entre le programme, l'enseignant et la classe, http://fr.allafrica.com/stories/200107050250.html, 2001.

FAUCQUEUR C., "La valorisation des innovations pédagogiques", *Économie et Management*, no. 119, CNDP, available at: http://www2.cndp.fr/revueEcoManagement/pdf/119/068073.pdf, April 2006.

FERRARI A., CACHIA R., PUNIE Y., Innovation and Creativity in Education and Training in the EU Member States, European Union, Brussels, 2009.

FEYFANT A., A la recherche de l'autonomie des établissements, Report, no. 118, 2017.

FILSER M., "Méthodologie d'élaboration d'une typologie de clientèle en marketing direct", *Revue Française du Marketing*, vol. 126, no. 1, pp. 57–66, 1990.

FOUGERES R., Mission Institut Carnot pour l'éducation, définition, expérimentation, essaimage, Final report, Ifé, 2016.

FREUD S., *Un souvenir d'enfance de Léonard de Vinci*, Gallimard, Paris, 1987.

GAGLIO G., *Sociologie de l'innovation*, PUF, Paris, 2014.

GARREL G., MOCK E., *La fabrique de l'innovation*, L'Harmattan, Paris, 2012.

GAUCHET M., *Le désenchantement du monde*, Gallimard, Paris, 1965.

GEFEN D., KARAHANNA E., STRAUB DETMAR W., "Trust and TAM in online shopping: an integrated model", *MIS Quarterly*, vol. 27, no. 1, 2003.

GIGLIO M., "5 dimensions pour étudier le changement de l'éducation du maître pour améliorer l'apprentissage créatif", *Journal for Educators, Teachers and Trainers*, vol. 5, no. 1, pp. 80–89, 2013.

GIGLIO M., PERRET-CLERMONT N., "Créer un objet nouveau en classe. Un dispositif d'innovation pédagogique et d'observation", *Éducation nouvelle*, pp. 211–237, 2016.

GIRARD R., *La voix méconnue du réel. Une théorie des mythes archaïques et modernes*, Grasset, Paris, 2002.

GOUDARZI K., La socialisation organisationnelle du client dans les entreprises de services, PhD Thesis, Aix-Marseille III University, 2005.

GOULET F., VINCK D., "L'innovation de retrait", *Revue française de sociologie*, vol. 53, no. 2, pp. 195–224, 2012.

GRAMSCI A., Prison Notebook, edited by R. PARIS, Gallimard, Paris, 1983.

GUIJARRO B.R., RAIMONDI G., *Estado del Arte Sobre las Innovaciones Educativas en America Latina*, Santiago, 2000.

GUILFORD J.P., *Characteristics of Creativity*, State Office of the Superintendent of Public Instruction, Gifted Children Section, Illinois, 1973.

HABERMAS J., *Théorie de l'agir communicationnel*, vol. 1, J.M. FERRY (trans.), Fayard, Paris, 1981.

HABERMAS J., *Théorie de l'agir communicationnel*, vol. 2, SCHEGEL J.L. (trans.), Fayard, Paris, 1987.

HAVELOCK R.G., Innovations in Education, strategies and tactics, working papers, British Journal of Educational Technology, Michigan, 1971.

HELLSTRÖM M., "School curriculum – is it worth trying?", in SAHLBERG P. (ed.), *International Conference on Curriculum Reform and Implementation in the 21st Century*, Istanbul, 2005.

HOUSSAYE J., *Le triangle pédagogique*, Peter Lang, Berne, 2000.

JOAS H., *La créativité de l'agir*, Éditions du Cerf, Paris, 1999.

KING S., *L'écriture: mémoires d'un métier*, Albin Michel, Paris, 2000.

KOSSELEK R., *Le futur passé. Contribution à la sémantique des temps historiques*, EHESS, Paris, 1990.

LAGREE M., *La bénédiction de Prométhée: religion et technologie*, Fayard, Paris, 1999.

LAVAL C., *L'école n'est pas une entreprise. Le néolibéralisme à l'assaut de l'enseignement public*, La Découverte, Paris, 2003.

LEROI-GOURHAN A., *L'homme et la machine. Évolution et techniques*, Albin Michel, Paris, 1943.

MARTIN S., Approcher le re-vécu d'une expérience graphique située en formation des enseignants, Conservatoire national des arts et métiers, Paris, 2015.

MILLER R., LOONEY J., SIEMENS G., Assessment competency: knowing what you know and learning analytics; it's time for a breakthrough Promethean Thinking Deep, Research Paper, Promethean Education Strategy Group, Blackburn, 2011.

MOSCOVICI S., *Psychologie des minorités actives*, PUF, Paris, 1991.

OBS, ""Le progrès est-il d'actualité?", *L'Obs*, no. 2731, p. 79, 9 March 2017.

PAUL M., *L'accompagnement : une posture professionnelle spécifique?*, L'Harmattan, Paris, 2004.

PAVE F., *La modernisation silencieuse des services publics*, L'Harmattan, Paris, 2006.

PELLETIER G., Piloter une innovation ou l'art de gérer l'inutile, Conférence, Deuxièmes journées du savoir vert, ministère de l'Agriculture et de la Pêche, Paris, 1994.

PROST A., *L'innovation*, intervention at the Biennale de l'éducation, Paris, 1996.

PROST A., *Du changement de l'école : les réformes de l'éducation de 1976 à nos jours*, Le Seuil, Paris, 2013.

PUOZZO I., "Pédagogie de la créativité : de l'émotion à l'apprentissage", *Éducation et socialisation*, vol. 33, available at: http://edso.revues.org/174, accessed 1 June 2017, 2013.

RADJOUR N., PRABHU J., *L'innovation frugale*, Diateino, Paris, 2015.

REGNIER C., *L'évaluation dans le système éducatif*, CNIRS, Paris, 2002.

REUTER Y. (ed.), Rapport sur les expérimentations liées à l'article 34 de la loi d'orientation et de programme pour l'école de 2005, report submitted to the French higher education council, June 2011.

RICŒUR P., *Temps et récit*, Le Seuil, Paris, 1984.

RICŒUR P., Identité narrative et communauté historique, *Cahier de Politique Autrement*, October 1994.

RICŒUR P., *La mémoire, l'histoire, l'oubli*, Le Seuil, Paris, 2003.

RIZZOLATTI D., CORRADO S., *Les neurones miroirs*, Odile Jacob, Paris, 2007.

ROGERS C.R., *A Way of Being*, Houghton Mifflin, Boston, 1995.

SATCHKOVA E., "L'éducation bulgare, du communisme à la démocratie : l'expérience d'une transition achevée", *Carrefours de l'éducation*, vol. 15, pp. 136–158, 2003.

SCHAEFFER J.M., *Pourquoi la fiction?*, Jacqueline Chambon, Paris, 1999.

SCHUMPETER J.A, A Theorical, Historical and Statistical Analysis of the Capitalist Process, *McGraw-Hill Book Company*, NY, Toronto, London, 1939.

SCHUMPETER J., *Capitalisme, socialisme et démocratie*, Payot, Paris, 1965.

SCHUTZ A., *Le chercheur et le quotidien: phénoménologie des sciences sociales*, Méridiens Klincksieck, Paris, 1987.

STERNBERG R., LUBART T., *Handbook of Creativity*, Cambridge University, 1999.

TADDÉI F. *et al.*, *Vers une société apprenante*, Report submitted to the French Minister for national and higher education, Paris, March 2017.

TORRANCE E.P., *Torrance Tests of Creative Thinking: Technical-norms Manual*, Research Edition, Princeton Press, Lexington, MA, 1966.

TROSA S., "Les réformes vues d'en bas", *Pyramides*, no. 17, available at: http://pyramides.revues.org/636, accessed 14 May 2017, 2009.

TROSA S., "L'État est-il encore régalien ou devient-il un réseau social?", *Pyramides*, no. 25, available at: http://pyramides.revues.org/967, accessed 13 May 2017, 2013.

VENKATECH *et al.*, "Age differences in technology adoption decisions: implications for a changing workforce", *Personnel Psychology*, no. 53, pp. 375–403, 2000.

VENKATESK, "Technology Acceptance in Education", *MIS Quaterly*, vol 2, no. 3, pp. 483–502, 2000–2003.

VIAL M., CAPARROS-MENCACCI N., *L'accompagnement professionnel? Méthode à l'usage des praticiens exerçant une fonction éducative*, De Boeck, Brussels, 2007.

VINCENT G. (ed.), *L'éducation prisonnière de la forme scolaire? Scolarisation et socialisation dans les sociétés industrielles*, PUF, Paris, 1994.

WAGNER P., *Sauver le progrès. Comment rendre l'avenir à nouveau désirable?*, La Découverte, Paris, 2016.

WATRELOT P. *et al.*, Innover pour une école plus juste et plus efficace, Synthèse des travaux du Conseil national de l'innovation et de la réussite éducative (Cniré), Report submitted to the French Minister for national and higher education, Paris, March 2017.

WATZLAWICK P., WEAKLAND J., FISCH R., *Changements: paradoxes et psychothérapie*, Le Seuil, Paris, 2014.

WEBER M., *L'éthique protestante et l'esprit du capitalisme*, Gallimard, Paris, 2004.

WINIKAMEN F., *Apprendre en imitant?*, PUF, Paris, 2015.

ZEKIAN S., "François Hartog et les temps de l'histoire", *Esprit*, no. 11, pp. 193–199, 2003.

# Index

## A, B, C



## D, E, F



## H, I, L

## N, O, P



## R, S, T



## V, W

Other titles from

iSTE

in

Innovation, Entrepreneurship and Management

## 2018

CORLOSQUET-HABART Marine, JANSSEN Jacques
*Big Data for Insurance Companies*
*(Big Data, Artificial Intelligence and Data Analysis Set – Volume 1)*

## 2017

AÏT-EL-HADJ Smaïl
*The Ongoing Technological System*
*(Smart Innovation Set – Volume 11)*

BAUDRY Marc, DUMONT Béatrice
*Patents: Prompting or Restricting Innovation?*
*(Smart Innovation Set – Volume 12)*

BÉRARD Céline, TEYSSIER Christine
*Risk Management: Lever for SME Development and Stakeholder Value Creation*

CHALENÇON Ludivine
*Location Strategies and Value Creation of International Mergers and Acquisitions*

CHAUVEL Danièle, BORZILLO Stefano
*The Innovative Company: An Ill-defined Object*
*(Innovation Between Risk and Reward Set – Volume 1)*

CORSI Patrick
*Going Past Limits To Growth*

D'ANDRIA Aude, GABARRET Inés
*Building 21st Century Entrepreneurship*
*(Innovation and Technology Set – Volume 2)*

DAIDJ Nabyla
*Cooperation, Coopetition and Innovation*
*(Innovation and Technology Set – Volume 3)*

FERNEZ-WALCH Sandrine
*The Multiple Facets of Innovation Project Management*
*(Innovation between Risk and Reward Set – Volume 4)*

FOREST Joëlle
*Creative Rationality and Innovation*
*(Smart Innovation Set – Volume 14)*

GUILHON Bernard
*Innovation and Production Ecosystems*
*(Innovation between Risk and Reward Set – Volume 2)*

HAMMOUDI Abdelhakim, DAIDJ Nabyla
*Game Theory Approach to Managerial Strategies and Value Creation*
*(Diverse and Global Perspectives on Value Creation Set – Volume 3)*

LALLEMENT Rémi
*Intellectual Property and Innovation Protection: New Practices and New Policy Issues*
*(Innovation between Risk and Reward Set – Volume 3)*

LAPERCHE Blandine
*Enterprise Knowledge Capital*
*(Smart Innovation Set – Volume 13)*

LEBERT Didier, EL YOUNSI Hafida
*International Specialization Dynamics*
*(Smart Innovation Set – Volume 9)*

MAESSCHALCK Marc
*Reflexive Governance for Research and Innovative Knowledge*
*(Responsible Research and Innovation Set – Volume 6)*

MASSOTTE Pierre
*Ethics in Social Networking and Business 1: Theory, Practice and Current Recommendations*
*Ethics in Social Networking and Business 2: The Future and Changing Paradigms*

MASSOTTE Pierre, CORSI Patrick
*Smart Decisions in Complex Systems*

MEDINA Mercedes, HERRERO Mónica, URGELLÉS Alicia
*Current and Emerging Issues in the Audiovisual Industry*
*(Diverse and Global Perspectives on Value Creation Set – Volume 1)*

MICHAUD Thomas
*Innovation, Between Science and Science Fiction*
*(Smart Innovation Set – Volume 10)*

PELLÉ Sophie
*Business, Innovation and Responsibility*
*(Responsible Research and Innovation Set – Volume 7)*

SAVIGNAC Emmanuelle
*The Gamification of Work: The Use of Games in the Workplace*

SUGAHARA Satoshi, DAIDJ Nabyla, USHIO Sumitaka
*Value Creation in Management Accounting and Strategic Management: An Integrated Approach*
*(Diverse and Global Perspectives on Value Creation Set –Volume 2)*

UZUNIDIS Dimitri, SAULAIS Pierre
*Innovation Engines: Entrepreneurs and Enterprises in a Turbulent World*
*(Innovation in Engineering and Technology Set – Volume 1)*

## 2016

BARBAROUX Pierre, ATTOUR Amel, SCHENK Eric
*Knowledge Management and Innovation*
*(Smart Innovation Set – Volume 6)*

BEN BOUHENI Faten, AMMI Chantal, LEVY Aldo
*Banking Governance, Performance And Risk-Taking: Conventional Banks Vs Islamic Banks*

BOUTILLIER Sophie, CARRE Denis, LEVRATTO Nadine
*Entrepreneurial Ecosystems (Smart Innovation Set – Volume 2)*

BOUTILLIER Sophie, UZUNIDIS Dimitri
*The Entrepreneur (Smart Innovation Set – Volume 8)*

BOUVARD Patricia, SUZANNE Hervé
*Collective Intelligence Development in Business*

GALLAUD Delphine, LAPERCHE Blandine
*Circular Economy, Industrial Ecology and Short Supply Chains*
*(Smart Innovation Set – Volume 4)*

GUERRIER Claudine
*Security and Privacy in the Digital Era*
*(Innovation and Technology Set – Volume 1)*

MEGHOUAR Hicham
*Corporate Takeover Targets*

MONINO Jean-Louis, SEDKAOUI Soraya
*Big Data, Open Data and Data Development*
*(Smart Innovation Set – Volume 3)*

MOREL Laure, LE ROUX Serge
*Fab Labs: Innovative User*
*(Smart Innovation Set – Volume 5)*

PICARD Fabienne, TANGUY Corinne
*Innovations and Techno-ecological Transition*
*(Smart Innovation Set – Volume 7)*

## 2015

CASADELLA Vanessa, LIU Zeting, DIMITRI Uzunidis
*Innovation Capabilities and Economic Development in Open Economies (Smart Innovation Set – Volume 1)*

CORSI Patrick, MORIN Dominique
*Sequencing Apple's DNA*

CORSI Patrick, NEAU Erwan
*Innovation Capability Maturity Model*

FAIVRE-TAVIGNOT Bénédicte
*Social Business and Base of the Pyramid*

GODÉ Cécile
*Team Coordination in Extreme Environments*

MAILLARD Pierre
*Competitive Quality and Innovation*

MASSOTTE Pierre, CORSI Patrick
*Operationalizing Sustainability*

MASSOTTE Pierre, CORSI Patrick
*Sustainability Calling*

## 2014

DUBÉ Jean, LEGROS Diègo
*Spatial Econometrics Using Microdata*

LESCA Humbert, LESCA Nicolas
*Strategic Decisions and Weak Signals*

## 2013

HABART-CORLOSQUET Marine, JANSSEN Jacques, MANCA Raimondo
*VaR Methodology for Non-Gaussian Finance*

## 2012

DAL PONT Jean-Pierre
*Process Engineering and Industrial Management*

MAILLARD Pierre
*Competitive Quality Strategies*

POMEROL Jean-Charles
*Decision-Making and Action*

SZYLAR Christian
*UCITS Handbook*

## 2011

LESCA Nicolas
*Environmental Scanning and Sustainable Development*

LESCA Nicolas, LESCA Humbert
*Weak Signals for Strategic Intelligence: Anticipation Tool for Managers*

MERCIER-LAURENT Eunika
*Innovation Ecosystems*

## 2010

SZYLAR Christian
*Risk Management under UCITS III/IV*

## 2009

COHEN Corine
*Business Intelligence*

ZANINETTI Jean-Marc
*Sustainable Development in the USA*

## 2008

CORSI Patrick, DULIEU Mike
*The Marketing of Technology Intensive Products and Services*

DZEVER Sam, JAUSSAUD Jacques, ANDREOSSO Bernadette
*Evolving Corporate Structures and Cultures in Asia / Impact of Globalization*

## 2007

AMMI Chantal
*Global Consumer Behavior*

## 2006

BOUGHZALA Imed, ERMINE Jean-Louis
*Trends in Enterprise Knowledge Management*

CORSI Patrick *et al.*
*Innovation Engineering: the Power of Intangible Networks*

Printed and bound by CPI Group (UK) Ltd, Croydon, CR0 4YY

16/09/2026

14963075-0002